Dynamical Systems

Shlomo Sternberg
Professor of Mathematics
Harvard University

Dover Publications, Inc.
Mineola, New York

Bibliographical Note

Dynamical Systems is a new work, first published by Dover Publications, Inc., in 2010.

International Standard Book Number

ISBN-13: 978-0-486-47705-3
ISBN-10: 0-486-47705-3

Manufactured in the United States by Courier Corporation
47705302
www.doverpublications.com

PREFACE

I have taught an advanced one semester undergraduate course here at Harvard on dynamical systems eight times in the past twenty years. This book represents how I usually teach this course. Over the preceding thirty years, that is from 1959 to 1989, I taught a research level graduate course in dynamical systems five times plus a graduate course in celestial mechanics twice. So as much thought had to go into what to omit from the course, as had to go into what to include.

Of course, over the past fifty years, there have been many advances in the field of dynamical systems, and some of them are presented in this book. But the major advances that impact on teaching such a course come from outside the field of dynamical systems, indeed from outside of mathematics.

The advent of the computer, especially the personal computer, together with mathematical platforms such as MATLAB or Mathematica allow for numerical and graphical implementations that were inconceivable fifty years ago. The ability to experiment and play with these implementations is an integral part of modern instruction in this field. For this reason the web site for this course contains a collection of MATLAB routines and a reference to the web site of Lynch for more programs.

Another major technological change is the advent of the web. Not only can many books and articles be accessed, but there are a variety of stand alone applets which can be run on the personal computer. For example, "pplane", by John Polking, is an excellent source for the study of phase portraits of autonomous systems of differential equations in the plane, and we use it here at Harvard for many of our mathematics classes. In the web site of this course we list a few relevant applets, but it is clearly impossible to keep up to date with a rapidly expanding resource.

Many of the figures in the book were drawn with MATLAB or with pplane, but some were downloaded from the web. For example, the figure on page 21 - showing the basins of attraction for Newton's method applied to finding the cube root of 1- was downloaded from the web. This picture appears on many web sites, and, unfortunately, I can not recall which of these web sites I downloaded it from. So my apologies for the lack of attribution.

All the photographs of mathematicians appearing in the book were downloaded from the web site http://www-history.mcs.st-andrews.ac.uk/ created by John J. O'Connor and Edmund F. Robertson.

The web site for the course, hosted both by Dover and by my Harvard web site, also contains slides of lectures which might be useful to someone teaching a course on this material. Due to scheduling restrictions I had to teach the course in two one and one half hour lectures per week. So each lecture on the web site should correspond to two one hour lectures.

The background that I assume as prerequisites for the course are a good knowledge of multivariate calculus and a thorough knowledge of linear algebra.

Here is a description of the various chapters, the table of contents gives additional details:

In chapter 1 I introduce the notion of iteration and fixed point behavior of map defined on an interval of the real line or on a subset of Euclidean n-dimensional space. The behavior under iteration of such a discrete map *is* the main subject of the course. Fifty years ago, the main emphasis of a course on dynamical systems would have been on asymptotic behavior of solutions of ordinary differential equations. The main result in this chapter is an analysis of Newton's method. This is probably the hardest chapter in the book from a technical point of view, but my feeling was that students shopping around for a course should not fooled by too gentle an approach at the beginning. From a purely logical standpoint, the course should start with the contraction fixed point theorem of Chapter 6. I then treat local behavior near a fixed point. I end this chapter with the remarkable prediction of Cayley as to the "fractal behavior" (to use Mandelbrot's term) of Newton's method in the complex plane. I also give a gentle introduction to renormalization group methods.

Chapter 2 discusses the bifurcation behavior of the logistic map, and ends with a crude description of Feigenbaum renormalization. In teaching an undergraduate course one would like to avoid, as much as possible, the phrase "it can be shown". But it is not always possible. I had to include a description of Feigenbaum's famous work, but could not present the technical details at an undergraduate level. Fortunately, I found (via the web) a beautiful description by Prof. Coppersmith of this work, which I present here.

Chapter 3 discusses, in technical detail, some properties of maps of an interval: a version of Sarkovky's theorem which guarantees the existence of periodic points of all periods, once there is a point of period three, and Singer's theorem which explains why we can not see most of them.

Chapter 4 introduces the important concept of conjugacy - when do we want to consider two dynamical systems as "the same"? It also gives the definition of "chaos" due to Devaney, J. Banks et al and proves that the logistic transformation L_4 is chaotic in their sense. For better or for worse, the term "chaos" has become faddish in the general public. The chapter ends with a gentle introduction to symbolic dynamics, a topic treated in more detail in Chapter 12.

Chapter 5 is an introduction to the statistical properties of dynamical systems. Thus, although the logistic transformation is chaotic, it has a very regular behavior from the point of view of statistical dynamics - it has an invariant density given by the arc sine law. Since this is an undergraduate course, I allow myself the freedom to meander. Having introduced the arc sine law in conjunction with L_4, I proceed to discuss the arc sine law in terms of its main application - fluctuation theory. I follow the classic treatment by Feller.

Chapter 6 discusses the contraction fixed point theorem, and gives several of its standard applications including the implicit function theorem (again) and the existence theorem for ordinary differential equations. As mentioned above, from a purely logical point of view, this should be the first chapter in the book.

Chapter 7 discusses the Hausdorff metric and Hutchinson's theorem, which is an application of the contraction fixed point theorem. The point of view here is that of my good friend Benoit Mandelbrot. A key philosophical point is that certain sets, such as the Cantor set or the Sierpinski gasket, which looked incredibly complicated at the beginning of the twentieth century, become remarkably simple if you have a platform (such as MATLAB) which includes iteration. I do not believe that this philosophical point has been fully digested in the applied sciences, despite the popularity of "chaos theory".

Chapter 8 discusses the concept of hyperbolicity, in particular the behavior of a transformation near a hyperbolic fixed point. This is a central concept in the modern theory of dynamical systems. It is also an old love of mine - many of my papers in the period 1954-1959 were devoted to this subject.

Chapter 9 is devoted to the Perron-Frobenius theorem and many of its myriad applications. Despite the fact that I assume a thorough knowledge of linear algebra as a prerequisite for this course, most of the students will not have seen this theorem before. For some reason that I do not understand, this important theorem has disappeared from the linear algebra curriculum here at Harvard. The applications include the Leslie model of population growth, the Google page ranking, and Fisher's answer to the question "why do we age?".

Chapter 10 contains some fragmental remains of what an old course in dynamical systems looked like in the distant past. Included are Lagrange's method of variation of constants, the Poincaré-Bendixon theorem about asymptotic behavior of differential equations in the plane and a study of the van der Pol and Lienard equations.

Chapter 11 is devoted to the Lotka-Volterra equations about the "struggle for life" and its modern developments, including their applications to replicator dynamics and evolutionary stable states. Having used the concept of entropy in the proof of a theorem of Zeeman, once again I meander, this time into information theory, and prove some key theorems of Shannon.

Chapter 12 returns to pure mathematics, and discusses some of the basic theorems in symbolic dynamics. I end with an exposition by Knill of a theorem about the Henon map. This exposition ties together many of the ideas developed earlier in the book.

Shlomo Sternberg
August 14, 2009

Dynamical Systems

Contents

Chapter 1

Iteration and fixed points.

1.1 Square roots.

Perhaps the oldest algorithm in recorded history is the Babylonian algorithm (circa 2000BCE) for computing square roots: If we want to find the square root of a positive number a we start with some approximation, $x_0 > 0$ and then recursively define

$$x_{n+1} = \frac{1}{2}\left(x_n + \frac{a}{x_n}\right). \tag{1.1}$$

This is a very effective algorithm which converges extremely rapidly.

Here is an illustration. Suppose we want to find the square root of 2 and start with the really stupid approximation $x_0 = 99$. We get:

99.00000000000000
49.51010101010101
24.77524840365297
12.42798706655775
6.29445708659966
3.30609848017316
1.95552056875300
1.48913306969968
1.41609819333465
1.41421481646475
1.41421356237365
1.41421356237309
1.41421356237309

1.1.1 Analyzing the steps.

For the first seven steps we are approximately dividing by two in passing from one step to the next, also (approximately) cutting the error - the deviation from the true value - in half.

After line eight the accuracy improves dramatically: the ninth value, 1.416... is correct to two decimal places. The tenth value is correct to five decimal places, and the eleventh value is correct to eleven decimal places.

To see why this algorithm works so well (for general $a > 0$), first observe that the algorithm is well defined, in that we are steadily taking the average of positive quantities, and hence, by induction, $x_n > 0$ for all n. Introduce the **relative error** in the n–th approximation:

$$e_n := \frac{x_n - \sqrt{a}}{\sqrt{a}}$$

so

$$x_n = (1 + e_n)\sqrt{a}.$$

As $x_n > 0$, it follows that

$$e_n > -1.$$

Then

$$x_{n+1} = \sqrt{a}\frac{1}{2}(1 + e_n + \frac{1}{1 + e_n}) = \sqrt{a}(1 + \frac{1}{2}\frac{e_n^2}{1 + e_n}).$$

This gives us a recursion formula for the relative error:

$$e_{n+1} = \frac{e_n^2}{2 + 2e_n}. \tag{1.2}$$

This implies that $e_{n+1} > 0$ so after the first step we are always overshooting the mark. Now $2e_n < 2 + 2e_n$ for $n \geq 1$ so (1.2) implies that

$$e_{n+1} < \frac{1}{2}e_n$$

so the error is cut in half (at least) at each stage after the first, and hence, in particular,

$$x_1 > x_2 > \cdots,$$

the iterates are steadily decreasing.

Eventually we will reach the stage that

$$e_n < 1.$$

From this point on, we use the inequality $2 + 2e_n > 2$ in (1.2) and we get the estimate

$$e_{n+1} < \frac{1}{2}e_n^2. \tag{1.3}$$

So if we renumber our approximation so that $0 \leq e_0 < 1$ then (ignoring the $1/2$ factor in (1.3)) we have

$$0 \leq e_n < e_0^{2^n}, \tag{1.4}$$

an exponential rate of convergence.

If we had started with an $x_0 < 0$ then all the iterates would be < 0 and we would get exponential convergence to $-\sqrt{a}$. Of course, had we been so foolish as to pick $x_0 = 0$ we could not get the iteration started.

1.2 Newton's method.

This is a generalization of the above algorithm to find the zeros of a function $P = P(x)$ and which reduces to (1.1) when $P(x) = x^2 - a$. It is

$$x_{n+1} = x_n - \frac{P(x_n)}{P'(x_n)}. \tag{1.5}$$

If we take $P(x) = x^2 - a$ then $P'(x) = 2x$ the expression on the right in (1.5) is

$$\frac{1}{2}\left(x_n + \frac{a}{x_n}\right)$$

so (1.5) reduces to (1.1).

Here is a graphic illustration of Newton's method applied to the function $y = x^3 - x$ with the initial point 2. Notice that what we are doing is taking the tangent to the curve at the point (x, y) and then taking as our next point, the intersection of this tangent with the x-axis. This makes the method easy to remember.

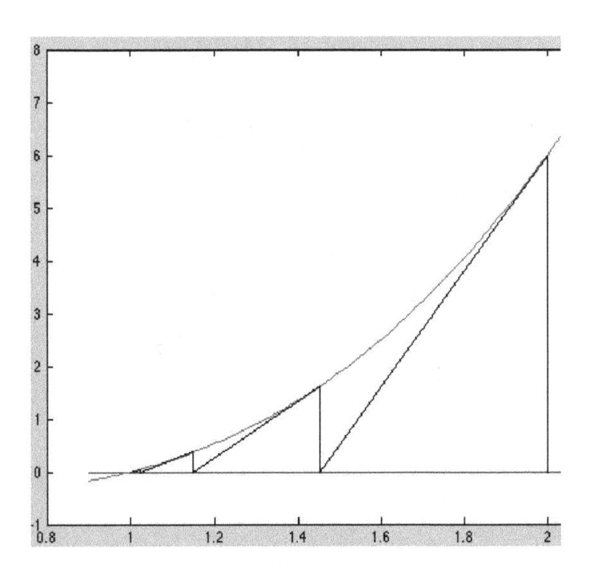

Caveat: In the general case we can not expect that "most" points will converge to a zero of P as was the case in the square root algorithm. After all, P might not have *any* zeros. Nevertheless, we will show in this section that if we are "close enough" to a zero - that $P(x_0)$ is "sufficiently small" in a sense to be made precise - then (1.5) converges exponentially fast to a zero.

1.2.1 A fixed point of the iteration scheme is a solution to our problem.

Notice that if x is a "fixed point" of this iteration scheme, i.e. if

$$x = x - \frac{P(x)}{P'(x)}$$

then $P(x) = 0$ and we have a solution to our problem. To the extent that x_{n+1} is "close to" x_n we will be close to a solution (the degree of closeness depending on the size of $P(x_n)$).

1.2.2 The guts of the method.

Before embarking on the formal proof, let us describe what is going on, on the assumption that we know the existence of a zero - say by graphically plotting the function. So let z be a zero for the function f of a real variable, and let x be a point in the interval $(z - \mu, z + \mu)$ of radius μ about z. Then

$$-f(x) = f(z) - f(x) = \int_x^z f'(s)ds$$

so

$$-f(x) - (z - x)f'(x) = \int_x^z (f'(s) - f'(x))ds.$$

Assuming $f'(x) \neq 0$ we may divide both sides by $f'(x)$ to obtain

$$\left(x - \frac{f(x)}{f'(x)}\right) - z = \frac{1}{f'(x)} \int_x^z (f'(s) - f'(x))ds. \tag{1.6}$$

Assume that for all $y \in (z - \mu, z + \mu)$ we have

$$|f'(y)| \geq \rho > 0 \tag{1.7}$$

$$|f'(y_1) - f'(y_2)| \leq \delta |y_1 - y_2| \tag{1.8}$$

$$\mu \leq \rho / \delta. \tag{1.9}$$

Then setting $x = x_{old}$ in (1.6) and letting

$$x_{new} := x - \frac{f(x)}{f'(x)}$$

in (1.6) we obtain

$$|x_{new} - z| \leq \frac{\delta}{\rho} \int_{x_{old}}^z |s - x_{old}| ds = \frac{\delta}{2\rho} |x_{old} - z|^2.$$

Since $|x_{old} - z| < \mu$ it follows that

$$|x_{new} - z| \leq \frac{1}{2}\mu$$

by (1.9). Thus the iteration

$$x \mapsto x - \frac{f(x)}{f'(x)} \tag{1.10}$$

is well defined. At each stage it more than halves the distance to the zero and has the quadratic convergence property

$$|x_{new} - z| \leq \frac{\delta}{2\rho}|x_{old} - z|^2.$$

The above argument was posited on the assumption that there is a zero z of f and that certain additional hypotheses were satisfied. But f might not have any zeros. Even if it does, unless some such stringent hypotheses are satisfied, there is no guarantee that the process will converge to the nearest root, or converge at all. Furthermore, encoding a computation for $f'(x)$ may be difficult. In practice, one replaces f' by an approximation, and only allows Newton's method to proceed if in fact it does not take us out of the interval. We will return to these points, but first rephrase the above argument in terms of a vector variable.

1.2.3 A vector version.

Now let f a function of a vector variable, with a zero at z and x a point in the ball of radius μ centered at z. Let $v_x := z - x$ and consider the function

$$t :\mapsto f(x + tv_x)$$

which takes the value $f(z)$ when $t = 1$ and the value $f(x)$ when $t = 0$. Differentiating with respect to t using the chain rule gives $f'(x + tv_x)v_x$ (where f' denotes the derivative =(the Jacobian matrix) of f. Hence

$$-f(x) = f(z) - f(x) = \int_0^1 f'(x + tv_x)v_x dt.$$

This gives

$$-f(x) - f'(x)v_x = -f(x) - f'(x)(z - x) = \int_0^1 [f'(x + tv_x) - f'(x)]v_x dt.$$

Applying $[f'(x)]^{-1}$ (which we assume to exist) gives the analogue of (1.6):

$$\left(x - [f'(x)]^{-1}f(x)\right) - z = [f'(x)]^{-1} \int_0^1 [f'(x + tv_x) - f'(x)]v_x dt.$$

Assume that
$$\|[f'(y)]^{-1}\| \leq \rho^{-1} \tag{1.11}$$
$$\|f'(y_1) - f'(y_2)\| \leq \delta\|y_1 - y_2\| \tag{1.12}$$

for all y, y_1, y_2 in the ball of radius μ about z, and assume also that $\mu \leq \rho/\delta$ holds. Setting $x_{old} = x$ and

$$x_{new} := x_{old} - [f'(x_{old})]^{-1} f(x_{old})$$

gives

$$\|x_{new} - z\| \leq \frac{\delta}{\rho} \int_0^1 t\|v_x\|\|v_x\| dt = \frac{\delta}{2\rho}\|x_{old} - z\|^2.$$

From here on we can argue as in the one dimensional case.

1.2.4 Problems with the implementation of Newton's method.

We return to the one dimensional case.

In numerical practice we have to deal with two problems: it may not be easy to encode the derivative, and we may not be able to tell in advance whether the conditions for Newton's method to work are indeed fulfilled.

In case f is a polynomial, MATLAB has an efficient command "polyder" for computing the derivative of f. Otherwise we replace the derivative by the slope of the secant, which requires the input of two initial values, call them x_- and x_c and replaces the derivative in Newton's method by

$$f'_{app}(x_c) = \frac{f(x_c) - f(x_-)}{x_c - x_-}.$$

$$f'_{app}(x_c) = \frac{f(x_c) - f(x_-)}{x_c - x_-}.$$

So at each stage of the Newton iteration we carry along two values of x, the "current value" denoted say by "xc" and the "old value" denoted by "x_". We also carry along two values of f, the value of f at xc denoted by fc and the value of f at x_ denoted by f_. So the Newton iteration will look like

```
fpc=(fc-f_)/(xc-x_);
xnew=xc-fc/fpc;
x_-=xc; f_=fc;
xc=xnew; fc=feval(fname,xc);
```

In the last line, the command feval is the MATLAB evaluation of a function command: if fname is a "script" (that is an expression enclosed in ' ') giving the name of a function, then feval(fname,x) evaluates the function at the point x.

The second issue - that of deciding whether Newton's method should be used at all - is handled as follows: If the zero in question is a critical point, so that $f'(z) = 0$, there is no chance of Newton's method working. So let us assume that $f'(z) \neq 0$, which means that f changes sign at z, a fact that we can verify by looking at the graph of f. So assume that we have found an interval $[a, b]$

containing the zero we are looking for, and such that f takes on opposite signs at the end-points:

$$f(a)f(b) < 0.$$

A sure but slow method of narrowing in on a zero of f contained in this interval is the "bisection method": evaluate f at the midpoint $\frac{1}{2}(a+b)$. If this value has a sign opposite to that of $f(a)$ replace b by $\frac{1}{2}(a+b)$. Otherwise replace a by $\frac{1}{2}(a+b)$. This produces an interval of half the length of $[a,b]$ containing a zero.

The idea now is to check at each stage whether Newton's method leaves us in the interval, in which case we apply it, or else we apply the bisection method.

We now turn to the more difficult existence problem.

1.2.5 The existence theorem.

For the purposes of the proof, in order to simplify the notation, let us assume that we have "shifted our coordinates" so as to take $x_0 = 0$. Also let

$$B = \{x : |x| \leq 1\}.$$

We need to assume that $P'(x)$ is nowhere zero, and that $P''(x)$ is bounded. In fact, we assume that there is a constant K such that

$$|P'(x)^{-1}| \leq K, \quad |P''(x)| \leq K, \quad \forall x \in B. \tag{1.13}$$

Proposition 1.2.1. *Let* $\tau = \frac{3}{2}$ *and choose the* K *in* (1.13) *so that* $K \geq 2^{3/4}$. *Let*

$$c = \frac{8}{3}\ln K.$$

Then if

$$|P(0)| \leq K^{-5} \tag{1.14}$$

the recursion (1.5) starting with $x_0 = 0$ *satisfies*

$$x_n \in B \quad \forall n \tag{1.15}$$

and

$$|x_n - x_{n-1}| \leq e^{-c\tau^n}. \tag{1.16}$$

In particular, the sequence $\{x_n\}$ *converges to a zero of* P.

We will prove a somewhat more general result: We will let τ be any real number satisfying

$$1 < \tau < 2$$

and we will choose c in terms of K and τ to make the proof work. First of all we notice that (1.15) is a consequence of (1.16) if c is sufficiently large. In fact,

$$x_j = (x_j - x_{j-1}) + \cdots + (x_1 - x_0)$$

so

$$|x_j| \leq |x_j - x_{j-1}| + \cdots + |x_1 - x_0|.$$

Using (1.16) for each term on the right gives

$$|x_j| \leq \sum_1^j e^{-c\tau^n} < \sum_1^\infty e^{-c\tau^n} < \sum_1^\infty e^{-cn(\tau-1)} = \frac{e^{-c(\tau-1)}}{1-e^{-c(\tau-1)}}.$$

Here the third inequality follows from writing $\tau = 1+(\tau-1)$ so by the binomial formula

$$\tau^n = 1 + n(\tau - 1) + \cdots > n(\tau - 1)$$

since $\tau > 1$. The equality is obtained by summing the geometric series.

We have shown that

$$|x_j| \leq \frac{e^{-c(\tau-1)}}{1-e^{-c(\tau-1)}}.$$

So if we choose c sufficiently large that

$$\frac{e^{-c(\tau-1)}}{1-e^{-c(\tau-1)}} \leq 1, \tag{1.17}$$

then (1.15) follows from (1.16).

This choice of c is conditioned by our choice of τ. But at least we now know that if we can arrange that (1.16) holds, then by choosing a possibly larger value of c (so that (1.16) continues to hold) we can guarantee that the algorithm keeps going.

So let us try to prove (1.16) by induction. If we assume it is true for n, we may write

$$|x_{n+1} - x_n| = |S_n P(x_n)|$$

where we set

$$S_n = P'(x_n)^{-1}. \tag{1.18}$$

We use the first inequality in (1.13) which says that

$$|P'(x)^{-1}| \leq K,$$

and the definition (1.5) for the case $n-1$ (which says that $x_n = x_{n-1} - S_{n-1}P(x_{n-1})$) to get

$$|S_n P(x_n)| \leq K|P(x_{n-1} - S_{n-1}P(x_{n-1}))|. \tag{1.19}$$

Taylor's formula with remainder says that for any twice continuously differentiable function f,

$$f(y+h) = f(y) + f'(y)h + R(y,h) \quad \text{where } |R(y,h)| \leq \frac{1}{2}\sup_z |f''(z)|h^2$$

where the supremum is taken over the interval between y and $y+h$. If we use Taylor's formula with remainder with

$$f = P,\ y = P(x_{n-1}), \text{and } -h = S_{n-1}P(x_{n-1}) = x_n - x_{n-1}$$

and the second inequality in (1.13) to estimate the second derivative, we obtain

$$|P(x_{n-1} - S_{n-1}P(x_{n-1}))|$$

$$\leq |P(x_{n-1}) - P'(x_{n-1})S_{n-1}P(x_{n-1})| + K|x_n - x_{n-1}|^2.$$

Substituting this inequality into (1.19), we get

$$|x_{n+1} - x_n| \leq K|P(x_{n-1}) - P'(x_{n-1})S_{n-1}P(x_{n-1})| + K^2|x_n - x_{n-1}|^2. \quad (1.20)$$

Now since $S_{n-1} = P'(x_{n-1})^{-1}$ the first term on the right vanishes and we get

$$|x_{n+1} - x_n| \leq K^2|x_n - x_{n-1}|^2 \leq K^2 e^{-2c\tau^n}.$$

Choosing c so that the induction works.

So in order to pass from n to $n+1$ in (1.16) we must have

$$K^2 e^{-2c\tau^n} \leq e^{-c\tau^{n+1}}$$

or

$$K^2 \leq e^{c(2-\tau)\tau^n}. \quad (1.21)$$

Since $1 < \tau < 2$ we can arrange for this last inequality to hold for $n = 1$ and hence for all n if we choose c sufficiently large.

Getting started.

To get started, we must verify (1.16) for $n = 1$ This says

$$S_0 P(0) \leq e^{-c\tau}$$

or

$$|P(0)| \leq \frac{e^{-c\tau}}{K}. \quad (1.22)$$

So we have proved:

Theorem 1.2.1. *Suppose that* (1.13) *holds and we have chosen K and c so that* (1.17) *and* (1.21) *hold. Then if $P(0)$ satisfies* (1.22) *the Newton iteration scheme converges exponentially to a zero of P in the sense that* (1.16) *holds.*

If we choose $\tau = \frac{3}{2}$ as in the proposition, let c be given by $K^2 = e^{3c/4}$ so that (1.21) just holds. This is our choice in the proposition. The inequality $K \geq 2^{3/4}$ implies that $e^{3c/4} \geq 4^{3/4}$ or

$$e^c \geq 4.$$

This implies that

$$e^{-c/2} \leq \frac{1}{2}$$

so (1.17) holds. Then

$$e^{-c\tau} = e^{-3c/2} = K^{-4}$$

so (1.22) becomes $|P(0)| \leq K^{-5}$ completing the proof of the proposition.

1.2.6 Review.

We have put in all the gory details, but it is worth reviewing the argument, and seeing how things differ from the special case of finding the square root. Our algorithm is

$$x_{n+1} = x_n - S_n[P(x_n)] \tag{1.23}$$

where S_n is chosen as (1.18). Taylor's formula gave (1.20) and with the choice (1.18) we get

$$|x_{n+1} - x_n| \leq K^2|x_n - x_{n-1}|^2. \tag{1.24}$$

In contrast to (1.4) we do not know that $K \leq 1$ so, once we get going, we can't quite conclude that the error vanishes as

$$r^{\tau^n}, \quad 0 < r < 1$$

with $\tau = 2$. But we can arrange that we eventually have such exponential convergence with any $\tau < 2$.

1.2.7 Basins of attraction.

The more decisive difference has to do with the "basins of attraction" of the solutions. For the square root, starting with any positive number ends us up with the positive square root. This was the effect of the $e_{n+1} < \frac{1}{2}e_n$ argument which eventually gets us to the region where the exponential convergence takes over. Every negative number leads us to the negative square root. So the "basin of attraction" of the positive square root is the entire positive half axis, and the "basin of attraction" of the negative square root is the entire negative half axis. The only "bad" point belonging to no basin of attraction is the point 0.

Even for cubic polynomials the *global* behavior of Newton's method is extraordinarily complicated. For example, consider the polynomial

$$P(x) = x^3 - x,$$

with roots at 0 and ± 1. We have

$$x - \frac{P(x)}{P'(x)} = x - \frac{x^3 - x}{3x^2 - 1} = \frac{2x^3}{3x^2 - 1}$$

so Newton's method in this case says to set

$$x_{n+1} = \frac{2x_n^3}{3x_n^2 - 1}. \tag{1.25}$$

There are obvious "bad" points where we can't get started, due to the vanishing of the denominator, $P'(x)$. These are the points $x = \pm\sqrt{1/3}$. These two points are the analogues of the point 0 in the square root algorithm.
We know from the general theory, that any point sufficiently close to 1 will converge to 1 under Newton's method and similarly for the other two roots, 0 and -1.

If $x > 1$, then $2x^3 > 3x^2 - 1$ since both sides agree at $x = 1$ and the left side is increasing faster, as its derivative is $6x^2$ while the derivative of the right hand side is only $6x$. This implies that if we start to the right of $x = 1$ we will stay to the right. The same argument shows that

$$2x^3 < 3x^3 - x$$

for $x > 1$. This is the same as

$$\frac{2x^3}{3x^2 - 1} < x,$$

which implies that if we start with $x_0 > 1$ we have $x_0 > x_1 > x_2 > \cdots$ and eventually we will reach the region where the exponential convergence takes over. So every point to the right of $x = 1$ is in the basin of attraction of the root $x = 1$. By symmetry, every point to the left of $x = -1$ will converge to -1.

But let us examine what happens in the interval $-1 < x_0 < 1$. For example, suppose we start with $x_0 = -\frac{1}{2}$. Then one application of Newton's method gives

$$x_1 = \frac{-.25}{3 \times .25 - 1} = 1.$$

In other words, one application of Newton's method lands us on the root $x = 1$, right on the nose. Notice that although $-.5$ is halfway between the roots -1 and 0, we land on the farther root $x = 1$. In fact, by continuity, if we start with x_0 close to $-.5$, then x_1 must be close to 1. So all points, x_0, sufficiently close to $-.5$ will have x_1 in the region where exponential convergence to $x = 1$ takes over. In other words, the basin of attraction of $x = 1$ will include points to the immediate left of $-.5$, even though -1 is the closest root.

Here are the results of applying Newton's method to the three close points

0.4472 , 0.4475 and 0.4480 with ten iterations:

0.4472	0.4475	0.4480
−0.4471	−0.4489	−0.4520
0.4467	0.4577	0.4769
−0.4443	−0.5162	−0.6827
0.4301	1.3699	−1.5980
−0.3576	1.1105	−1.2253
0.1483	1.0146	−1.0500
−0.0070	1.0003	−1.0034
0.0000	1.0000	−1.0000
−0.0000	1.0000	−1.0000
0.0000	1.0000	−1.0000

Periodic points.

Suppose we have a point x which satisfies

$$\frac{2x^3}{3x^2-1} = -x.$$

So one application of Newton's method lands us at $-x$, and a second lands us back at x. The above equation is the same as

$$0 = 5x^3 - x = x(5x^2 - 1)$$

which has roots, $x = 0, \pm\sqrt{1/5}$. So the points $\pm\sqrt{1/5}$ form a cycle of order two: Newton's method cycles between these two points and hence does not converge to any root. In fact, in the interval $(-1, 1)$ there are infinitely many points that don't converge to any root. We will return to a description of this complicated type of phenomenon later.

1.2.8 Cayley's complex version

If we apply Newton's method to cubic or higher degree polynomials and to complex numbers instead of real numbers, the results are even more spectacular. This phenomenon was first discovered by Cayley, and was published in a short article which appeared in the second issue of the American Journal of Mathematics in 1879. After describing Newton's method, Cayley writes, concerning a polynomial with roots A,B,C... in the complex plane:

> The problem is to determine the regions of the plane such that P, taken at pleasure anywhere within one region, we arrive ultimately at the point A, anywhere within another region we arrive at the point B, and so for the several points representing the root of the equation.
>
> The solution is easy and elegant for the case of a quadric equation; but the next succeeding case of a cubic equation appears to present considerable difficulty.

This paper of Cayley's was the starting point for many future investigations.

With the advent of computers, we can see how complicated the problem really is. The next figure shows, via color coding, the regions corresponding to the three roots of 1, i.e. the results of applying Newton's method to the polynomial $x^3 - 1$. The roots themselves are indicated by the + signs.

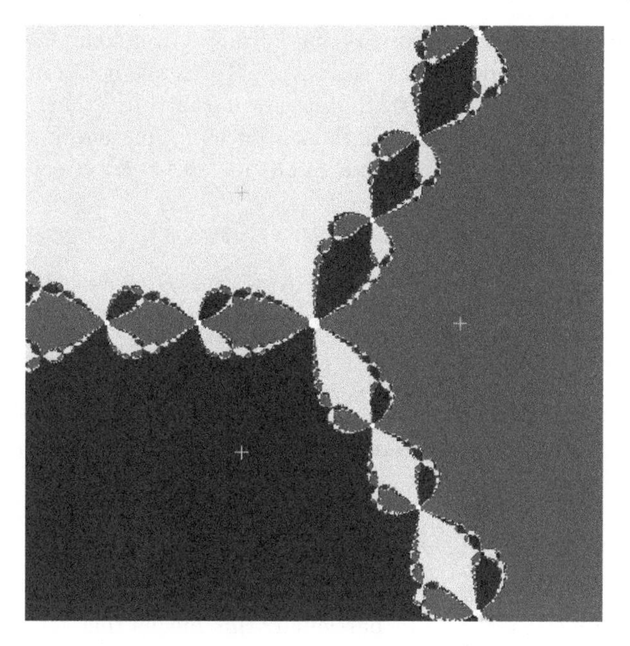

Here is a picture of the great man:

Arthur Cayley (August 16, 1821 - January 26, 1895)

1.3 The implicit function theorem via Newton's method.

Let us return to the positive aspect of Newton's method. You might ask, how can we ever guarantee in advance that an inequality such as (1.14) holds? The answer comes from considering not a single function, P, but rather a parameterized family of functions: Suppose that u ranges over some interval, or more generally, over some region in a vector space. To fix the notation, suppose that this region contains the origin, $\mathbf{0}$. Suppose that P is a function of u and x, and depends continuously on (u, x). Suppose that as a function of x, the function P is twice differentiable and satisfies (1.13) for all values of u (with the same fixed K).

$$\left|\left(\frac{\partial P}{\partial x}\right)^{-1}\right| \leq K, \quad \left|\frac{\partial^2 P}{\partial (x)^2}\right| \leq K, \quad \forall x \in B, \ u. \qquad (1.13)$$

Finally, suppose that

$$P(\mathbf{0}, 0) = 0. \qquad (1.26)$$

Then the continuity of P guarantees that for $|u|$ and $|x_0|$ sufficiently small, the condition (1.14) holds, that is

$$|P(u, x_0)| < r$$

where r is small enough to guarantee that x_0 is in the basin of attraction of a zero of the function $P(u, \cdot)$ In particular, this means that for $|u|$ sufficiently small, we can find an $\epsilon > 0$ such that all x_0 satisfying $|x_0| < \epsilon$ are in the basin of attraction of the same zero of $P(u, \cdot)$. By choosing a smaller neighborhood, given say by $|u| < \delta$, starting with $x_0 = 0$ and applying Newton's method to $P(u, \cdot)$, we obtain a sequence of x values which converges exponentially to a solution of

$$P(u, x) = 0. \qquad (1.27)$$

satisfying

$$|x| < \epsilon.$$

Furthermore, starting with any x_0 satisfying $|x_0| < \epsilon$ we also get exponential convergence to the same zero. In particular, there can not be two distinct solutions to (1.27) satisfying $|x| < \epsilon$, since starting Newton's method at a zero gives (inductively) $x_n = x_0$ for all n. Thus we have constructed a unique function

$$x = g(u)$$

satisfying

$$P(u, g(u)) \equiv 0. \qquad (1.28)$$

This is the guts of the *implicit function theorem*. We have proved it under assumptions which involve the second derivative of P which are not necessary for the truth of the theorem. (We will remedy this later in the book.) However these stronger assumptions that we have made *do* guarantee exponential convergence of our algorithm.

1.3.1 The continuity, differentiability of the implicit function, and the computation of its derivative.

For the sake of completeness, we discuss the basic properties of the function g given by the implicit function theorem: its continuity, differentiability, and the computation of its derivative.

Continuity.

We wish to prove that g is continuous at any point u in a neighborhood of $\mathbf{0}$. This means: given $\beta > 0$ we can find $\alpha > 0$ such that

$$|h| < \alpha \Rightarrow |g(u+h) - g(u)| < \beta. \tag{1.29}$$

We know that this is true at $u = 0$, where we could choose any $\epsilon' > 0$ at will, and then conclude that there is a $\delta' > 0$ with $|g(u)| < \epsilon'$ if $|u| < \delta'$.

To prove (1.29) at a general point, just choose $(u, g(u))$ instead of $(\mathbf{0}, 0)$ as the origin of our coordinates, and apply the preceding results to this new data.

We obtain a solution f to the equation $P(u+h, f(u+h)) = 0$ with $f(u) = g(u)$ which is continuous at $h = 0$. In particular, for $|h|$ sufficiently small, we will have $|u+h| \leq \delta$, and $|f(u+h)| < \epsilon$, our original ϵ and δ in the definition of g. The uniqueness of the solution to our original equation then implies that $f(u+h) = g(u+h)$, proving (1.29).

Differentiability.

Suppose that P is continuously differentiable with respect to all variables. We have

$$0 \equiv P(u+h, g(u+h)) - P(u, g(u))$$

so, by the definition of the derivative,

$$0 = \frac{\partial P}{\partial u} h + \frac{\partial P}{\partial x}[g(u+h) - g(u)] + o(h) + o[g(u+h) - g(u)].$$

If u is a vector variable, say $\in \mathbb{R}^n$, then $\frac{\partial P}{\partial u}$ is a matrix. The terminology $o(s)$ means some expression which approaches zero so that $o(s)/s \to 0$. So

$$g(u+h) - g(u) = -\left[\frac{\partial P}{\partial x}\right]^{-1}\left[\frac{\partial P}{\partial u}\right] h - o(h) - \left[\frac{\partial P}{\partial x}\right]^{-1} o[g(u+h) - g(u)]. \tag{1.30}$$

As a first pass through this equation, observe that by the continuity that we have already proved, we know that $[g(u+h) - g(u)] \to 0$ as $h \to 0$. The expression $o([g(u+h)-g(u)])$ is, by definition of o, smaller than any constant times $|g(u+h)-g(u)|$ provided that $|g(u+h)-g(u)|$ itself is sufficiently small. This means that for sufficiently small $[g(u+h)-g(u)]$ we have

$$|o[g(u+h)-g(u)]| \leq \frac{1}{2K}|g(u+h)-g(u)|$$

where we may choose K so that

$$\left|\left[\frac{\partial P}{\partial x}\right]^{-1}\right| \leq K.$$

So bringing the last term in (1.30) over to the other side gives

$$|g(u+h)-g(u)| - \frac{1}{2}|g(u+h)-g(u)| \leq \left| \left[\frac{\partial P}{\partial x}\right]^{-1} \left[\frac{\partial P}{\partial u}\right] h\right| + o(|h|),$$

and we get an estimate of the form

$$|g(u+h)-g(u)| \leq M|h|$$

for some suitable constant, M. So the term $o[g(u+h)-g(u)]$ becomes $o(h)$. Plugging this back into our equation (1.30) shows that g is differentiable with

$$\frac{\partial g}{\partial u} = -\left[\frac{\partial P}{\partial x}\right]^{-1}\left[\frac{\partial P}{\partial u}\right]. \tag{1.31}$$

Statement of the theorem.

To summarize, the the version of the implicit function theorem that we have proved says:

Theorem 1.3.1. The implicit function theorem. *Let $P = P(u, x)$ be a differentiable function with $P(0,0) = 0$ and $\left[\frac{\partial P}{\partial x}\right](0,0)$ invertible. Then there exist $\delta > 0$ and $\epsilon > 0$ such that $P(u,x) = 0$ has a unique solution with $|x| < \epsilon$ for each $|u| < \delta$. This defines the function $x = g(u)$. The function g is differentiable and its derivative is given by (1.31).*

We have proved the theorem under more stringent hypotheses than necessary for the truth of the implicit function in order to get an exponential rate of convergence to the solution. We will provide the details of the more general version, as a consequence of the contraction fixed point theorem, later on. We should point out now, however, that nothing in our discussion of Newton's method or the implicit function theorem depended on x being a single real variable. The entire discussion goes through unchanged if x is a vector variable. Then $\partial P/\partial x$ is a matrix, and (1.31) must be understood as matrix multiplication. Similarly, the condition on the second derivative of p must be understood in terms of matrix norms. We will return to these points later.

For now we will give some interesting applications of the implicit function theorem to the problem of iteration of maps.

1.4 Attractors and repellers.

Over the next two chapters we will study the behavior of iterations of a map of an interval of the real line into the real line. But we will let this map depend on a parameter. So we will be studying the iteration (in x) of a function, F, of two real variables x and μ . We will need to make various hypothesis concerning the differentiability of F. We will always assume it is at least C^2 (has continuous partial derivatives up to the second order). We may also need C^3 in which case we explicitly state this hypothesis. We write

$$F_\mu(x) = F(x, \mu)$$

and will be interested in the change of behavior of F_μ as μ varies.

We begin by studying the case of a single map. In other words we are holding μ fixed. Here is some notation which we will be using: Let

$$f : X \to X$$

be a differentiable map where X is an interval on the real line.

1.4.1 Attractors.

A point $p \in X$ is called a **fixed point** if

$$f(p) = p.$$

A fixed point a is called an **attractor** or an **attractive** fixed point or a **stable** fixed point if

$$|f'(a)| < 1. \tag{1.32}$$

The reason for this terminology is that points sufficiently close to an attractive fixed point, a, converge to a geometrically upon iteration. Indeed,

$$f(x) - a = f(x) - f(a) = f'(a)(x - a) + o(x - a)$$

by the definition of the derivative. Hence taking $b < 1$ to be any number larger than $|f'(a)|$ then for $|x - a|$ sufficiently small, $|f(x) - a| \leq b|x - a|$. So starting with $x_0 = x$ and iterating $x_{n+1} = f(x_n)$ gives a sequence of points with $|x_n - a| \leq b^n|x - a|$.

1.4.2 The basin of attraction of an attractor.

The **basin of attraction** of an attractive fixed point is the set of all x such that the sequence $\{x_n\}$ converges to a where $x_0 = x$ and $x_{n+1} = f(x_n)$. Thus the basin of attraction of an attractive fixed point a will always include a neighborhood of a, but it may also include points far away, and may be a very complicated set as we saw in the example of Newton's method applied to a cubic.

1.4.3 Repellers.

A fixed point, r, is called a *repeller* or a *repelling* or an *unstable* fixed point if

$$|f'(r)| > 1. \tag{1.33}$$

Points near a repelling fixed point are pushed away upon iteration.

1.4.4 Superattractors.

An attractive fixed point s with

$$f'(s) = 0 \tag{1.34}$$

is called **superattractive** or **superstable**. Near a superstable fixed point, s, the iterates converge faster than any geometrical rate to s.

For example, in Newton's method,

$$f(x) = x - \frac{P(x)}{P'(x)}$$

so

$$f'(x) = 1 - \frac{P'(x)}{P'(x)} + \frac{P(x)P''(x)}{(P'(x)^2} = \frac{P(x)P''(x)}{P'(x)^2}.$$

So if a is a zero of P, then it is a superattractive fixed point.

1.4.5 Notation for iteration.

The notation $f^{\circ n}$ will mean the n-fold composition,

$$f^{\circ n} = f \circ f \circ \cdots \circ f \quad (n\text{times}).$$

1.4.6 Periodic points.

A fixed point of $f^{\circ n}$ is called a **periodic** point of *period* n . If p is a periodic point of period n, then so are each of the points

$$p,\ f(p),\ f^{\circ 2}(p),\ \ldots, f^{\circ(n-1)}(p)$$

and the chain rule says that at each of these points the derivative of $f^{\circ n}$ is the same and is given by

$$(f^{\circ n})'(p) = f'(p)f'(f(p)) \cdots f'(f^{\circ(n-1)}(p)).$$

If any one of these points is an attractive fixed point for f^n then so are all the others. We speak of an **attractive periodic orbit**. Similarly for repelling.

A periodic point will be superattractive for $f^{\circ n}$ if and only if at least one of the points $p, f(p), \ldots f^{\circ(n-1)}(p)$ satisfies $f('q) = 0$.

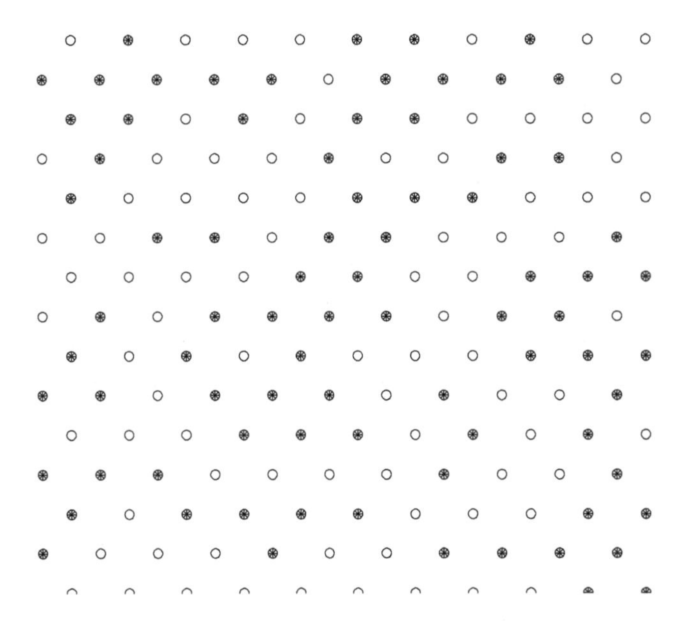

Figure 1.1: p=.2

1.5 Renormalization group

We illustrate these notions in an example: consider a hexagonal lattice in the plane. This means that each lattice point has six nearest neighbors. Let each site be occupied or not independently of the others with a common probability $0 \leq p \leq 1$ for occupation. In *percolation theory* the problem is to determine whether or not there is a positive probability for an infinitely large *cluster* of occupied sites. (By a cluster we mean a connected set of occupied sites.) We plot some figures with $p = .2$, $.5$, and $.8$ respectively. For problems such as this there is a *critical probability* p_c: for $p < p_c$ the probability of of an infinite cluster is zero, while it is positive for for $p > p_c$. One of the problems in percolation theory is to determine p_c for a given lattice.

For the case of the hexagonal lattice in the plane, it turns out that $p_c = \frac{1}{2}$. We won't prove that here, but arrive at the value $\frac{1}{2}$ as the solution to a problem which seems to be related to the critical probability problem in many cases. The idea of the *renormalization group* method is that many systems exhibit a similar behavior at different scales, a property known as *self similarity*. Understanding the transformation properties of this self similarity yields important information about the system. This is the goal of the *renormalization group method*. Rather than attempt a general definition, we use the hexagonal lattice as a first and elementary illustration:

Replace the original hexagonal lattice by a coarser hexagonal lattice as fol-

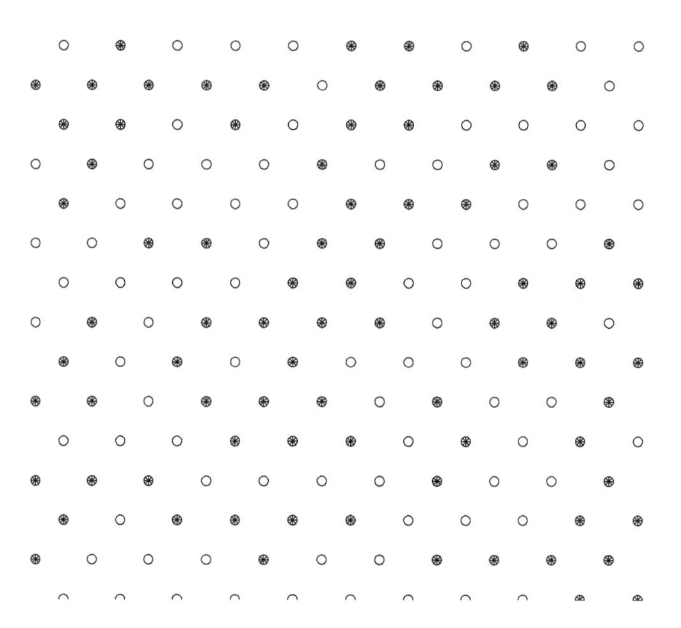

Figure 1.2: p=.5

Figure 1.3: p=.8

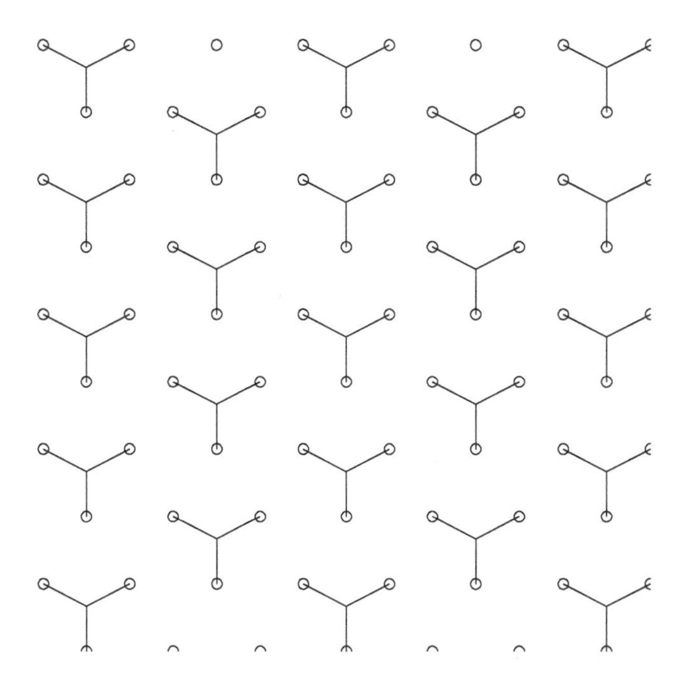

Figure 1.4: The original hexagonal lattice organized into groups of three adjacent vertices.

lows: pick three adjacent vertices on the original hexagonal lattice which form an equilateral triangle. This then organizes the lattice into a union of disjoint equilateral triangles, all pointing in the same direction, where, alternately, two adjacent lattice points on a row form a base of a triangle and the third lattice point is a vertex of a triangle from an adjacent row . The center of these triangles form a new (coarser) hexagonal lattice, in fact one where the distance between sites has been increased by a factor of three. See Figures 1.4 and 1.5.

Each point on our new hexagonal lattice is associated with exactly three points on our original lattice. Now assign a probability, p' to each point of our new lattice by the principle of majority rule: a new lattice point will be declared occupied if a majority of the associated points of the old lattice are occupied. Since our triangles are disjoint, these probabilities are independent. We can achieve a majority if all three sites are occupied (which occurs with probability p^3) or if two out of the three are occupied (which occurs with probability $p^2(1-p)$ with three choices as to which two sites are occupied). Thus

$$p' = p^3 + 3p^2(1-p). \tag{1.35}$$

This has three fixed points: $0, 1, \frac{1}{2}$. The derivative at $\frac{1}{2}$ is $\frac{3}{2} > 1$, so it is repelling. The points 0 and 1 are superattracting. So starting with any $p > \frac{1}{2}$, iteration leads rapidly towards the state where all sites are occupied, while starting with

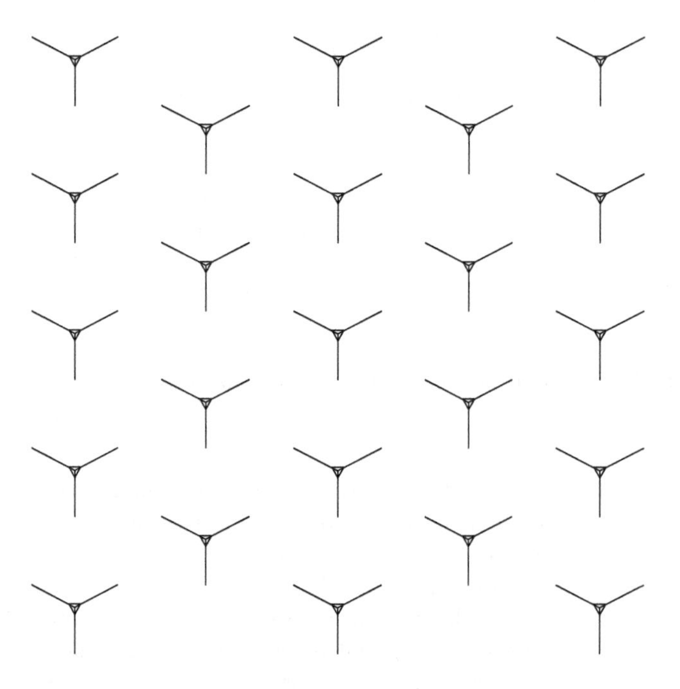

Figure 1.5: The new hexagonal lattice with edges emanating from each vertex, indicating the input for calculating p' from p.

$p < \frac{1}{2}$ leads rapidly under iteration towards the totally empty state. The point $\frac{1}{2}$ is an unstable fixed point for the renormalization transformation.

1.6 Iteration for kindergarten.

Suppose that we have drawn a graph of the map f, and have also drawn the x-axis and the diagonal line $y = x$. The iteration of f starting with an initial point x_0 on the x-axis can be visualized as follows:

- Draw the vertical line from x_0 until it hits the graph (at $(x_0, f(x_0))$).

- Draw the horizontal line to the diagonal (at $f(x_0), f(x_0))$.

- Call this new x value x_1, so $x_1 = f(x_0)$.

- Draw the vertical line to the graph (at $(x_1, f(x_1))$.

- Continue.

This method is known as **graphical iteration.**

We illustrate this for the graphical iteration of the quadratic map $f(x) = x^2 + .15$ starting with the initial point .75. The fixed points of f are obtained by solving the quadratic equation

$$x^2 - x + .15 = 0$$

and hence are given by

$$p_\pm = \frac{1}{2} \pm \sqrt{.1}.$$

The derivative of f is $2x$ so p_+ is an unstable fixed point while p_- is a stable fixed point.

Notice that we are moving away from the unstable fixed point, and as we continue then iteration we move towards the stable fixed point.

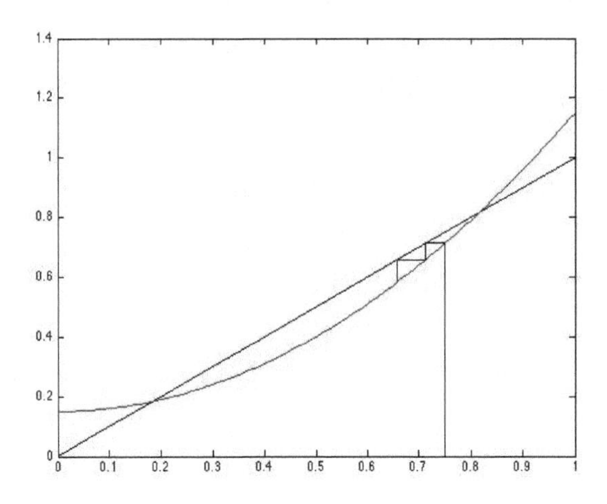

Figure 1.6: The first few steps.

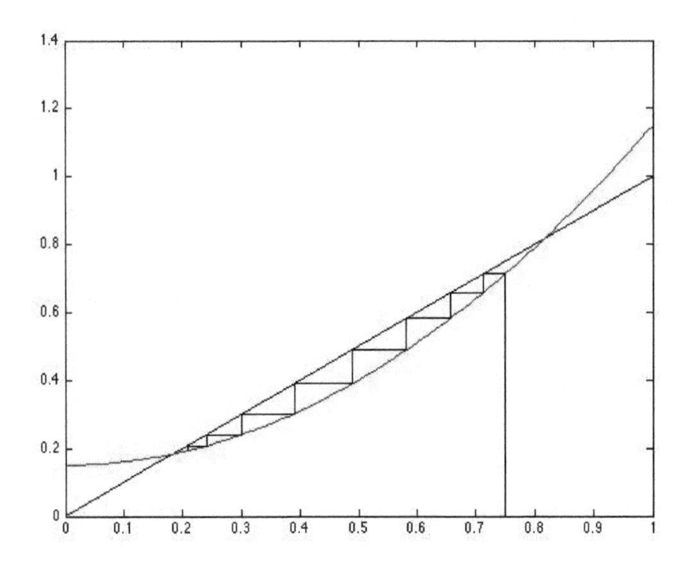

Figure 1.7: More iterations lead to the attractive fixed point.

Chapter 2

Bifurcations.

In this chapter we will study the behavior of iterations of a map of an interval of the real line into the real line. But we will let this map depend on a parameter. So we will be studying the iteration (in x) of a function, F, of two real variables x and μ . As mentioned above, we will need to make various hypothesis concerning the differentiability of F. We will always assume it is at least C^2 (has continuous partial derivatives up to the second order). We may also need C^3 in which case we explicitly state this hypothesis. We write

$$F_\mu(x) = F(x, \mu)$$

and are interested in the change of behavior of F_μ as μ varies. Before developing a general theory, we study a famous example.

2.1 The logistic family.

In population biology one considers iteration of the **logistic function**

$$L_\mu(x) := \mu x(1 - x). \tag{2.1}$$

Here $0 < \mu$ is a real parameter and x represents a proportion of a population, so we are mainly interested in $0 \leq x \leq 1$. For any fixed value of μ, the maximum of L_μ as a function of x is achieved at $x = \frac{1}{2}$ and the maximum value is $\frac{1}{4}\mu$. On the other hand, $L_\mu(x) \geq 0$ when $\mu \geq 0$ and $0 \leq x \leq 1$. So for any value of μ with $0 \leq \mu \leq 4$, the map

$$x \mapsto L_\mu(x)$$

maps the unit interval into itself. For $\mu > 4$, portions of $[0, 1]$ are mapped into the range $x > 1$. A second operation of L_μ maps these points to the range $x < 0$ and then are swept offto $-\infty$ under successive applications of L_μ. So for now, we will restrict attention to $0 \leq \mu \leq 4$. We will deal with $\mu > 4$ later.

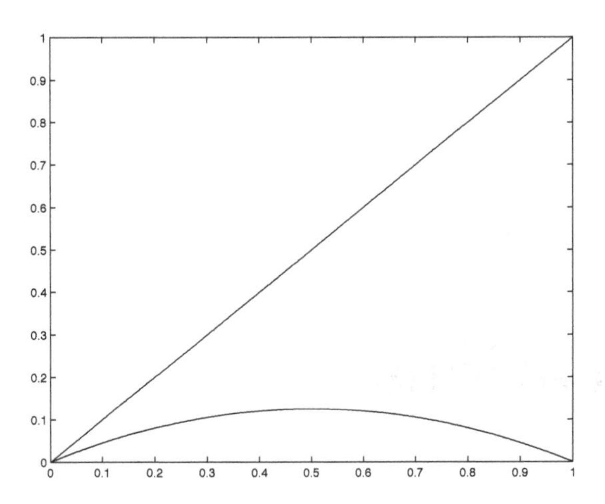

Figure 2.1: $\mu = .5$.

For any value of μ the fixed points of L_μ are 0 and $1 - \frac{1}{\mu}$. Since $L'_\mu(x) = \mu - 2\mu x$,

$$L'_\mu(0) = \mu, \quad L'_\mu(1 - \frac{1}{\mu}) = 2 - \mu. \tag{2.2}$$

2.1.1 $0 < \mu \leq 1$.

For $0 < \mu < 1$, 0 is the only fixed point of L_μ on $[0, 1]$ since the other fixed point, $1 - \frac{1}{\mu}$, is negative. On this range of μ, the point 0 is an attracting fixed point since $0 < L'_\mu(0) = \mu < 1$. Under iteration, all points of $[0, 1]$ tend to 0 under the iteration. The population "dies out".

2.1.2 $\mu = 1$.

For $\mu = 1$ we have

$$L_1(x) = x(1 - x) < x, \quad \forall x > 0.$$

Each successive application of L_1 to an $x \in (0, 1]$ decreases its value. The limit of the successive iterates can not be positive since 0 is the only fixed point. So all points in $(0, 1]$ tend to 0 under iteration, but ever so slowly, since $L'_1(0) = 1$. In fact, for $x < 0$, the iterates drift off to more negative values and then tend to $-\infty$.

2.1.3 $\mu > 1$.

For all $\mu > 1$, the fixed point, 0, is repelling, and the unique other fixed point, $1 - \frac{1}{\mu}$, lies in $[0, 1]$.

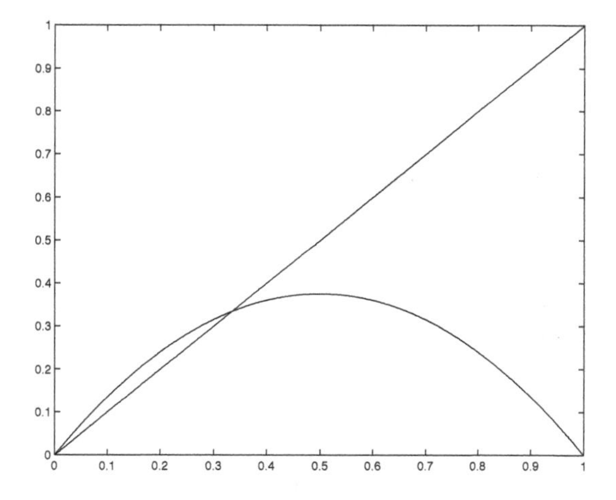

Figure 2.2: $\mu = 1.5$.

For $1 < \mu < 3$ we have

$$|L'_\mu(1 - \frac{1}{\mu})| = |2 - \mu| < 1,$$

so the non-zero fixed point is attractive.

We will see that the basin of attraction of $1 - \frac{1}{\mu}$ is the entire open interval $(0, 1)$, but the behavior is slightly different for the two domains, $1 < \mu \leq 2$ and $2 < \mu < 3$:

In the first of these ranges there is, eventually, a steady approach toward the fixed point from one side or the other; in the second, the iterates (eventually) bounce back and forth from one side to the other as they converge in towards the fixed point. The graphical iteration spirals in. Here are the details:

2.1.4 $1 < \mu < 2$.

For $1 < \mu < 2$ the non-zero fixed point lies between 0 and $\frac{1}{2}$ and the derivative at this fixed point is $2 - \mu$ and so lies between 1 and 0. Figure 2.2 gives the graph for $\mu = 1.5$:

Behavior when the initial point is $< 1 - \frac{1}{\mu}$.

Suppose that x lies between 0 and the fixed point, $1 - \frac{1}{\mu}$. For this range of x we have

$$\frac{1}{\mu} < 1 - x$$

so, multiplying by μx we get

$$x < \mu x(1 - x) = L_\mu(x).$$

Also, L_μ is monotone increasing for $0 < x < \frac{1}{2}$. So for $x < 1 - \frac{1}{\mu}$, $L_\mu(x) < L_\mu(1 - \frac{1}{\mu}) = 1 - \frac{1}{\mu}$. Thus the iterates steadily increase toward $1 - \frac{1}{\mu}$, eventually converging geometrically with a rate close to $2 - \mu$.

Behavior when the initial point satisfies $1 - \frac{1}{\mu} < x < \frac{1}{\mu}$.

If

$$1 - \frac{1}{\mu} < x$$

then $1 - x < \frac{1}{\mu}$ so, multiplying by μx gives

$$L_\mu(x) < x.$$

If, in addition,

$$x \leq \frac{1}{\mu}$$

then

$$L_\mu(x) \geq 1 - \frac{1}{\mu}.$$

To see this observe that the function L_μ has only one critical point, and that is a maximum. Since $L_\mu(1 - \frac{1}{\mu}) = L_\mu(\frac{1}{\mu}) = 1 - \frac{1}{\mu}$, we conclude that the minimum value is achieved at the end points of the interval $[1 - \frac{1}{\mu}, \frac{1}{\mu}]$.

So on the range $1 - \frac{1}{\mu} < x < \frac{1}{\mu}$ the iterates steadily decrease towards the fixed point, eventually converging to the fixed point at a geometric rate close to $2 - \mu$.

If $x = \frac{1}{\mu}$ then $L_\mu(x) = 1 - \frac{1}{\mu}$. So with one application of L_μ we hit the fixed point on the nose.

Behavior when the initial point satisfies $\frac{1}{\mu} < x < 1$.

On this range $0 < L_\mu(x) < 1 - \frac{1}{\mu}$, and then, after the first application of L_μ the iterates steadily increase toward the fixed point.

Of course, for any value of μ we have $L_\mu(1) = 0$, which is a fixed point (in our case unstable).

Summary.

So on the range $1 < \mu < 2$ the behavior of L_μ is as follows: All points $0 < x < 1 - \frac{1}{\mu}$ steadily increase toward the fixed point, $1 - \frac{1}{\mu}$. All points satisfying $1 - \frac{1}{\mu} < x < \frac{1}{\mu}$ steadily decrease toward the fixed point. The point $\frac{1}{\mu}$ satisfies $L_\mu(\frac{1}{\mu}) = 1 - \frac{1}{\mu}$ and so lands on the non-zero fixed point after one application. The points satisfying $\frac{1}{\mu} < x < 1$ get mapped by L_μ into the interval $0 < x < 1 - \frac{1}{\mu}$, In other words, they overshoot the mark, but then steadily increase towards the non-zero fixed point. Of course $L_\mu(1) = 0$ which is always true.

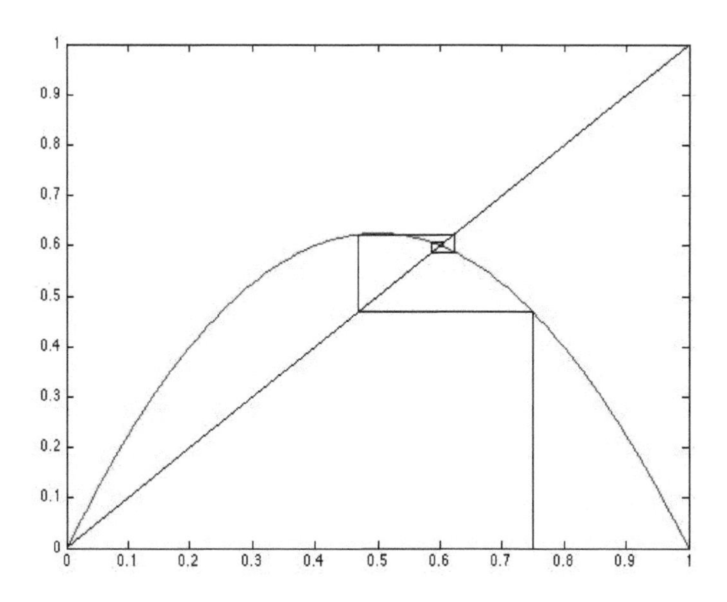

Figure 2.3: Graphical iteration of $L_{2.5}$ with initial point .75.

2.1.5 $\mu = 2$ - the fixed point is superattractive.

When $\mu = 2$, the points $\frac{1}{\mu}$ and $1-\frac{1}{\mu}$ coincide and equal $\frac{1}{2}$ with $L_2'(\frac{1}{2}) = 0$. There is no "steadily decreasing" region, and the fixed point, $\frac{1}{2}$ is superattractive - the iterates zoom into the fixed point faster than any geometrical rate.

2.1.6 $2 < \mu < 3$.

Here the fixed point $1 - \frac{1}{\mu} > \frac{1}{2}$ while $\frac{1}{\mu} < \frac{1}{2}$. The derivative at this fixed point is negative:

$$L_\mu'(1 - \frac{1}{\mu}) = 2 - \mu < 0.$$

So the fixed point $1 - \frac{1}{\mu}$ is an attractor, but as the iterates converge to the fixed points, they oscillate about it, alternating from one side to the other. The entire interval $(0, 1)$ is in the basin of attraction of the fixed point. To see this will take some work.

Before going into the details of the argument, we illustrate the result in Figure 2.3 via graphical iteration with $\mu = 2.5$ and initial point $x_0 = .75$.

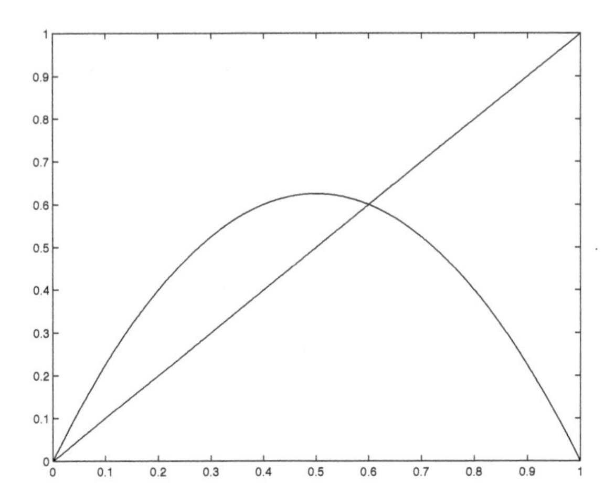

Figure 2.4: $\mu = 2.5,\ \frac{1}{\mu} = .4,\ 1 - \frac{1}{\mu} = .6.$

Proof that the entire open interval $(0, 1)$ is the basin of attraction of the fixed point $1 - \frac{1}{\mu}$.

The graph of L_μ lies entirely above the line $y = x$ on the interval $(0, 1 - \frac{1}{\mu}]$. In particular, it lies above the line $y = x$ on the subinterval $[\frac{1}{\mu}, 1 - \frac{1}{\mu}]$ and takes its maximum at $\frac{1}{2}$. So $\frac{\mu}{4} = L_\mu(\frac{1}{2}) > L_\mu(1 - \frac{1}{\mu}) = 1 - \frac{1}{\mu}$. Hence L_μ maps the interval $[\frac{1}{\mu}, 1 - \frac{1}{\mu}]$ onto the interval $[1 - \frac{1}{\mu}, \frac{\mu}{4}]$. The map L_μ is decreasing to the right of $\frac{1}{2}$, so it is certainly decreasing to the right of $1 - \frac{1}{\mu}$. Hence it maps the interval $[1 - \frac{1}{\mu}, \frac{\mu}{4}]$ into an interval whose right hand end point is $1 - \frac{1}{\mu}$ and whose left hand end point is $L_\mu(\frac{\mu}{4})$. We claim that

$$L_\mu(\frac{\mu}{4}) > \frac{1}{2}.$$

This amounts to showing that

$$\frac{\mu^2(4 - \mu)}{16} > \frac{1}{2}$$

or that

$$\mu^2(4 - \mu) > 8.$$

So we need only check the values of $\mu^2(4 - \mu)$ at the end points, 2 and 3, of the range of μ we are considering, where the values are 8 and 9.

So we have proved that the image of $[\frac{1}{\mu}, 1 - \frac{1}{\mu}]$ is the same as the image of $[\frac{1}{2}, 1 - \frac{1}{\mu}]$ and is $[1 - \frac{1}{\mu}, \frac{\mu}{4}]$. The image of this interval is the interval $[L_\mu(\frac{\mu}{4}), 1 - \frac{1}{\mu}]$, with $\frac{1}{2} < L_\mu(\frac{\mu}{4})$. If we apply L_μ to this interval, we get an interval to the right of $1 - \frac{1}{\mu}$ with right end point $L_\mu^2(\frac{\mu}{4}) < L_\mu(\frac{1}{2}) = \frac{\mu}{4}$. The image of the interval

$[1-\frac{1}{\mu}, L_\mu^2(\frac{\mu}{4})]$ must be strictly contained in the image of the interval $[1-\frac{1}{\mu}, \frac{\mu}{4}]$, and hence we conclude that

$$L_\mu^3(\frac{\mu}{4}) > L_\mu(\frac{\mu}{4}).$$

Continuing in this way we see that under even powers, the image of $[\frac{1}{2}, 1-\frac{1}{\mu}]$ is a sequence of nested intervals whose right hand end point is $1-\frac{1}{\mu}$ and whose left hand end points are

$$\frac{1}{2} < L_\mu(\frac{\mu}{4}) < L_\mu^3(\frac{\mu}{4}) < \cdots .$$

We claim that this sequence of points converges to the fixed point, $1-\frac{1}{\mu}$. If not, it would have to converge to a fixed point of L_μ^2 different from 0 and $1-\frac{1}{\mu}$. We shall show that there are no such points. Indeed, a fixed point of L_μ^2 is a zero of

$$L_\mu^2(x) - x = \mu L_\mu(x)(1 - L_\mu(x)) = \mu[\mu x(1-x)][1-\mu x(1-x)] - x.$$

Two roots of this quartic polynomial, are the fixed points of L_μ,which are 0 and $1-\frac{1}{\mu}$. So the quartic polynomial factors into a quadratic polynomial times $\mu x(x-1+\frac{1}{\mu})$. A check shows that this quadratic polynomial is

$$-\mu^2x^2 + (\mu^2+\mu)x - \mu - 1.$$

The $b^2 - 4ac$ for this quadratic function is

$$\mu^2(\mu^2 - 2\mu - 3) = \mu^2(\mu+1)(\mu-3) \tag{2.3}$$

which is negative for $2 < \mu < 3$ so the quadratic has no real roots. We thus conclude that the iterates of any point in $(\frac{1}{\mu}, \frac{\mu}{4}]$ oscillate about the fixed point, $1-\frac{1}{\mu}$ and converge in towards it, eventually with the geometric rate of convergence a bit less than $\mu - 2$. The graph of L_μ is strictly above the line $y = x$ on the interval $(0, \frac{1}{\mu}]$ and hence the iterates of L_μ are strictly increasing so long as they remain in this interval. Furthermore they can't stay there, for this would imply the existence of a fixed point in the interval and we know that there is none. Thus they eventually get mapped into the interval $[\frac{1}{\mu}, 1-\frac{1}{\mu}]$ and the oscillatory convergence takes over.

Finally, since L_μ is decreasing on $[1-\frac{1}{\mu}, 1]$, any point in $[1-\frac{1}{\mu}, 1)$ is mapped into $(0, 1-\frac{1}{\mu}]$ and so converges to the non-zero fixed point.

Summary

In short, every point in $(0, 1)$ is in the basin of attraction of the non-zero fixed point and (except for the points $\frac{1}{\mu}$ and the fixed point itself) eventually converge toward it in a "spiral" fashion.

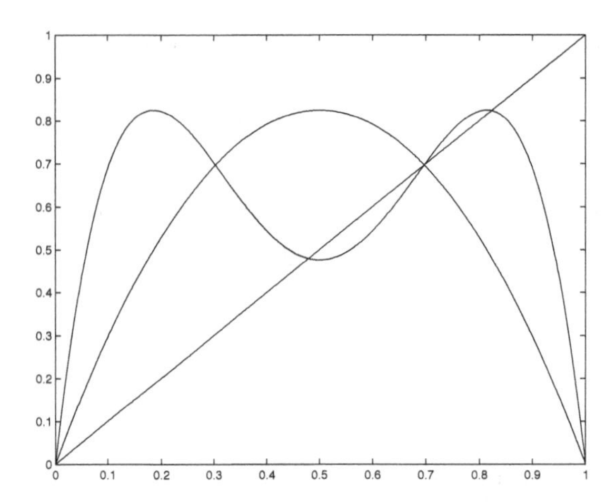

Figure 2.5: $\mu = 3.3$, graphs of $y = x,\ y = L_\mu(x),\ y = L)\mu^{(2)}(x)$. The graph of $L_\mu^{(2)}$ has four points of intersection with the line $y = x$: the two (repelling) fixed points of L_μ and two points of period two.

2.1.7 $\mu = 3$.

Much of the analysis of the preceding case applies here. The differences are: the quadratic equation (2.3) for seeking points of period two now has a (double) root. But this root is $\frac{2}{3} = 1 - \frac{1}{\mu}$ which is the fixed point. So there is still no point of period two other than the fixed points. The iterates continue to spiral in, but now ever so slowly since $L'_\mu(\frac{2}{3}) = -1$.

2.1.8 $\mu > 3$, points of period two appear.

For $\mu > 3$ we have

$$L'_\mu(1 - \frac{1}{\mu}) = 2 - \mu < -1$$

so both fixed points, 0 and $1 - \frac{1}{\mu}$ are repelling. But now (2.3) has two real roots which are

$$p_{2\pm} = \frac{1}{2} + \frac{1}{2\mu} \pm \frac{1}{2\mu}\sqrt{(\mu+1)(\mu-3)}.$$

Both points $p_\pm$ of period two lie in $(0, 1)$.

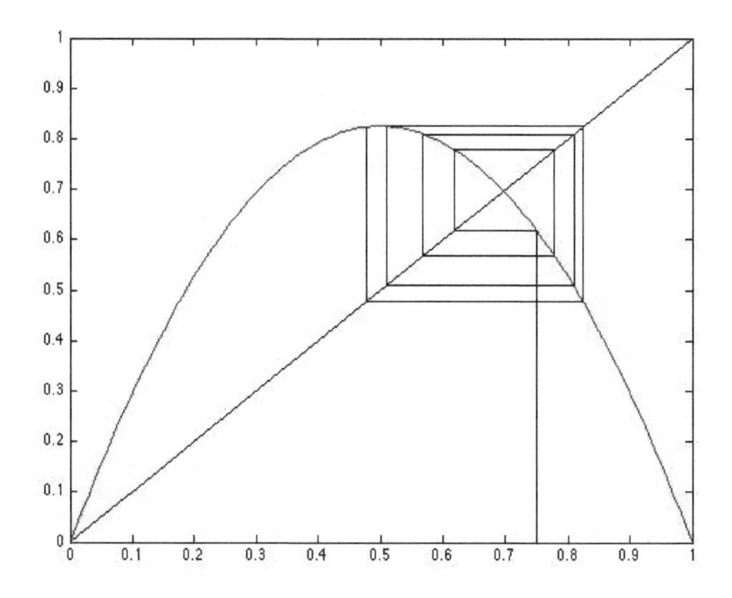

Figure 2.6: Graphical iteration for $\mu = 3.3$, nine steps.

The derivative of $L_\mu^{(2)}$ at these points of period two is given by

$$\begin{aligned}
(L_\mu^{(2)})'(p_{2\pm}) &= L'_\mu(p_{2+})L'_\mu(p_{2-}) \\
&= (\mu - 2\mu p_{2+})(\mu - 2\mu p_{2-}) \\
&= \mu^2 - 2\mu^2(p_{2+} + p_{2-}) + 4\mu^2 p_{2+}p_{2-} \\
&= \mu^2 - 2\mu^2(1 + \frac{1}{\mu}) + 4\mu^2 \times \frac{1}{\mu^2}(\mu + 1) \\
&= -\mu^2 + 2\mu + 4.
\end{aligned}$$

This last expression equals 1 when $\mu = 3$ as we already know. It decreases as μ increases reaching the value -1 when $\mu = 1 + \sqrt{6}$.

2.1.9 $3 < \mu < 1 + \sqrt{6}$.

In this range the fixed points are repelling and both period two points are attracting. There will be points whose images end up, after a finite number of iterations, on the non-zero fixed point. All other points in $(0, 1)$ are attracted to the period two cycle. We omit the proof.

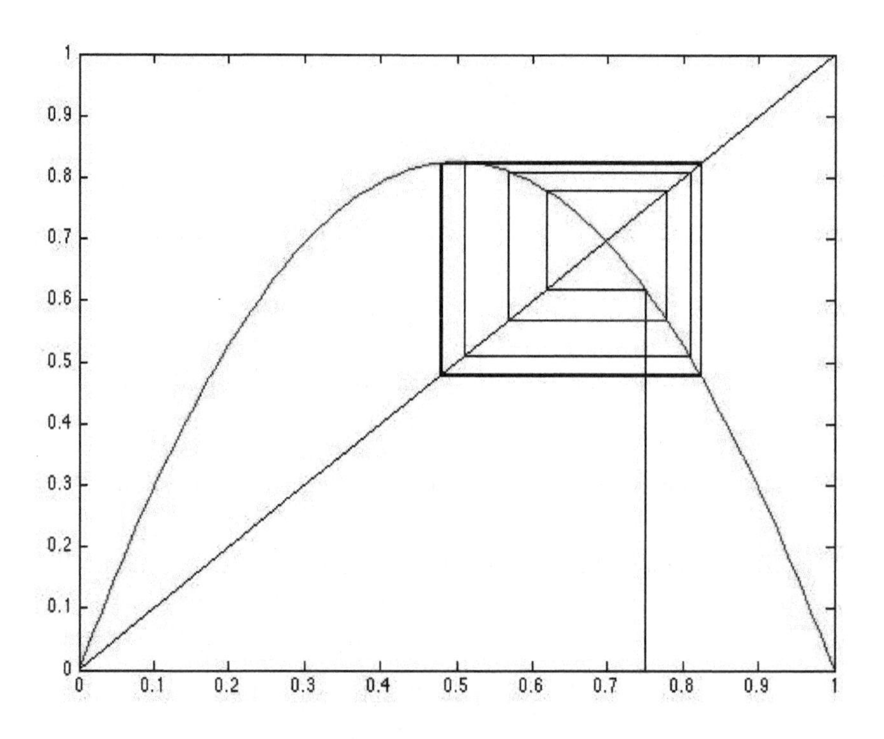

Figure 2.7: Graphical iteration for $\mu = 3.3$, twenty five steps.

2.1.10 Superattracting period two points.

Notice also that there is a unique value of μ in this range where

$$p_{2+}(\mu) = \frac{1}{2}.$$

Indeed, looking at the formula for p_{2+} we see that this amounts to the condition that $\sqrt{(\mu+1)(\mu-3)} = 1$ or

$$\mu^2 - 2\mu - 4 = 0.$$

The positive solution to this equation is given by $\mu = s_2$ where

$$s_2 = 1 + \sqrt{5}.$$

At s_2, the period two points are superattracting, since one of them coincides with $\frac{1}{2}$ which is the maximum of L_{s_2}.

2.1.11 $1 + \sqrt{6} < \mu$.

Once μ passes $1 + \sqrt{6} = 3.449499...$ the points of period two become unstable and (stable) points of period four appear. Initially these are stable, but as μ increases they become unstable (at the value $\mu = 3.544090...$) and bifurcate into period eight points, initially stable.

2.1.12 Reprise.

The total scenario so far, as μ increases from 0 to about 3.55, is as follows: For $\mu < b_1 := 1$, there is no non-zero fixed point. Past the first bifurcation point, $b_1 = 1$, the non-zero fixed point has appeared close to zero. When μ reaches the first superattractive value , $s_1 := 2$, the fixed point is at .5 and is superattractive. As μ increases, the fixed point continues to move to the right. Just after the second bifurcation point, $b_2 := 3$, the fixed point has become unstable and two stable points of period two appear, one to the right and one to the left of .5. The leftmost period two point moves to the right as we increase μ, and at $\mu = s_2 := 1 + \sqrt{5} = 3.23606797...$ the point .5 is a period two point, and so the period two points are superattractive. When μ passes the second bifurcation value $b_2 = 1 + \sqrt{6} = 3.449..$ the period two points have become repelling and attracting period four points appear.

In fact, this scenario continues. The period 2^{n-1} points appear at bifurcation values b_n. They are initially attracting, and become superattracting at $s_n > b_n$ and become unstable past the next bifurcation value $b_{n+1} > s_n$ when the period 2^n points appear.

Figure 2.9 gives the graph of the first four bifurcations:

Here is a MATLAB program for producing Figure 2.9, modified very slightly from Lynch.

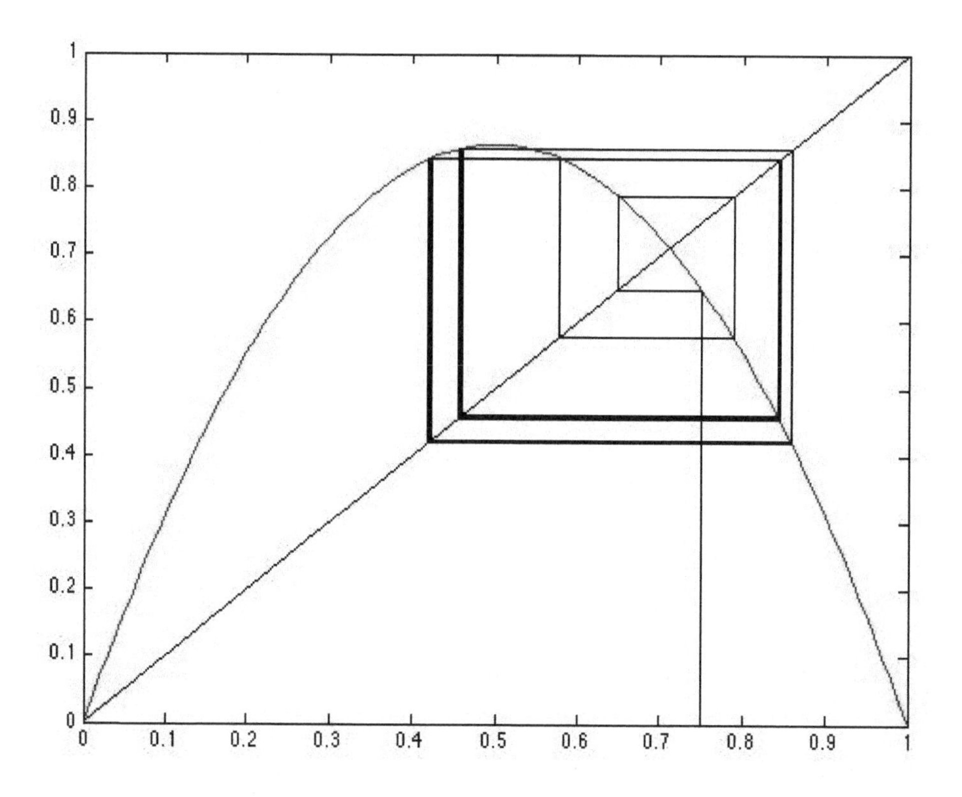

Figure 2.8: Graphical iteration for $\mu = 3.46$, twenty five steps. The attractive period four points become evident.

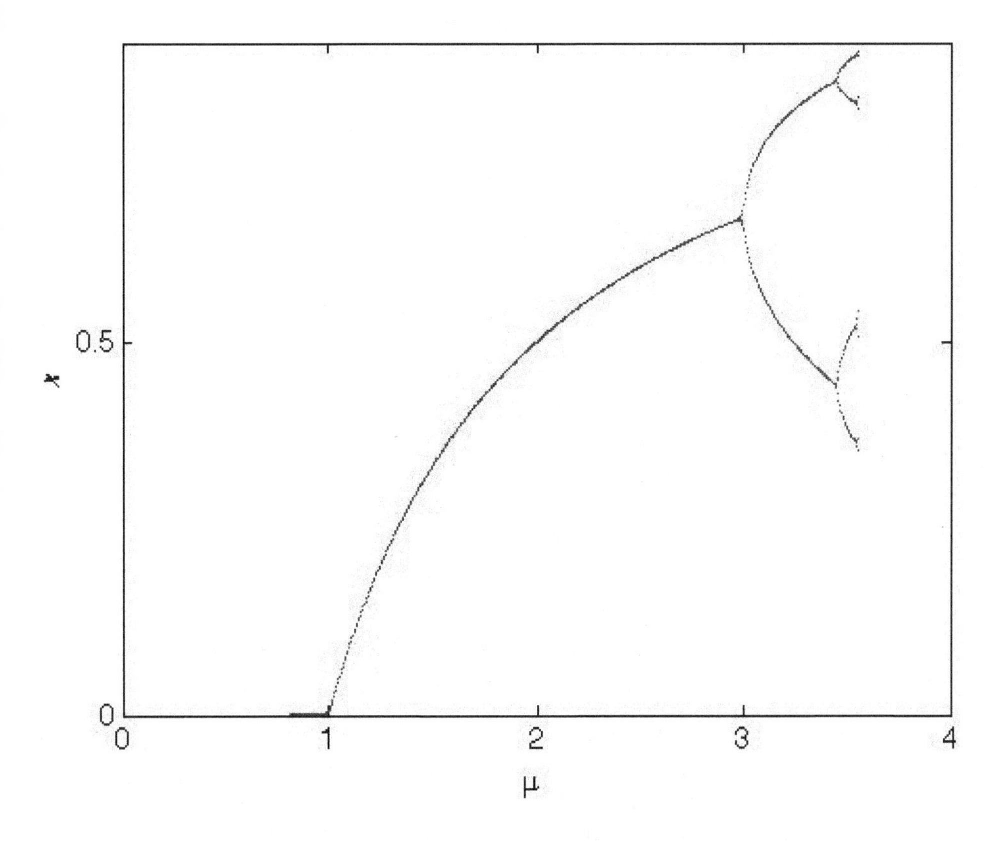

Figure 2.9: The graph of the first four bifurcations: For each value of r ranging in steps of .005 from 0 to 3.55 the values of $L_r^{\circ k}(x_0)$ were computed for 100 values of k (where x_0 was chosen as 0.4). Then only the last 30 values were kept, and these were plotted against r.

```
clear, itermax=100;
finalits=30;finits=itermax-(finalits-1);
for r=0:0.005:4
x=0.4; xo=x; for n=2:itermax
xn=r*xo*(1-xo);
x=[x xn];
xo=xn;
end
plot(r*ones(finalits),x(finits:itermax),'.','MarkerSize',1)
hold on
end
fsize=15; set(gca,'xtick',[0:1:4],'FontSize',fsize), set(gca,'ytick',[0,0.5,1],'FontSize',fsize
xlabel('mu','FontSize',fsize), ylabel('itx','FontSize',fsize), hold off
```

The (numerically computed) bifurcation points and superstable points are tabulated as:

n	b_n	s_n
1	1.000000	2.000000
2	3.000000	3.236068
3	3.449499	3.498562
4	3.544090	3.554641
5	3.564407	3.566667
6	3.568759	3.569244
7	3.569692	3.569793
8	3.569891	3.569913
9	3.569934	3.569946
∞	3.569946	3.569946

The values of the b_n are obtained by numerical experiment. Later, we shall describe a method for computing the s_n using Newton's method.

We should point out that this is still just the beginning of the story. For example, an attractive period three cycle appears at about 3.83. We shall come back to all of these points, but first go back and discuss theoretical problems associated to bifurcations, in particular, the "fold bifurcation" and the "period doubling bifurcation".

2.2 The fold bifurcation.

As mentioned, we will be studying the iteration (in x) of a function, F, of two real variables x and μ . To repeat once more: we will need to make various hypothesis concerning the differentiability of F. We will always assume it is at least C^2 (has continuous partial derivatives up to the second order). We may also need C^3 in which case we explicitly state this hypothesis. We write

$$F_\mu(x) = F(x, \mu)$$

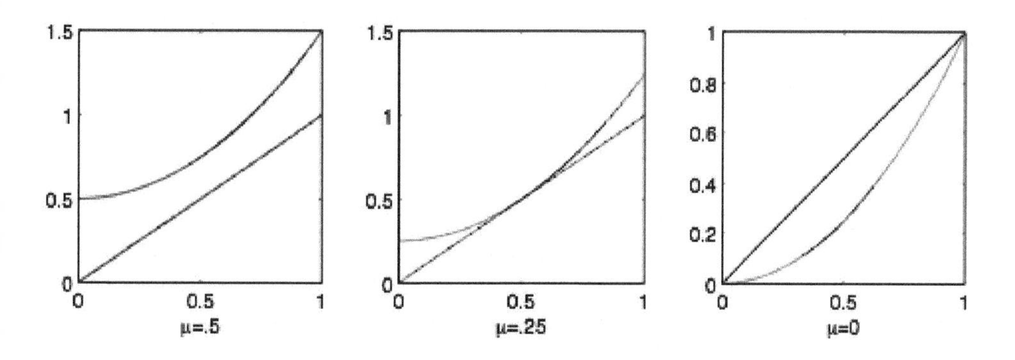

Figure 2.10: $y = x^2 + \mu$ for $\mu = .5,\ .25$ and 0.

and are interested in the change of behavior of F_μ as μ varies.

Before embarking on the study of bifurcations let us observe that if p is a fixed point of F_μ and $F'_\mu(p) \neq 1$, then for ν close to μ, the transformation F_ν has a unique fixed point close to p. Indeed, the implicit function theorem applies to the function

$$P(x,\nu) := F(x,\nu) - x$$

since

$$\frac{\partial P}{\partial x}(p,\mu) \neq 0$$

by hypothesis. We conclude that there is a curve of fixed points $x(\nu)$ with $x(\mu) = p$.

The first type of bifurcation we study is the **fold bifurcation** where there is no (local) fixed point on one side of the bifurcation value, b, where a fixed point p appears at $\mu = b$ with $F'_\mu(p) = 1$, and at the other side of b the map F_μ has two fixed points, one attracting and the other repelling.

As an example consider the quadratic family

$$Q(x,\mu) = Q_\mu(x) := x^2 + \mu.$$

Fixed points must be solutions of the quadratic equation

$$x^2 - x + \mu = 0,$$

whose roots are

$$p_\pm = \frac{1}{2} \pm \frac{1}{2}\sqrt{1-4\mu}.$$

For

$$\mu > b = \frac{1}{4}$$

these roots are not real. The parabola $x^2 + \mu$ lies entirely above the line $y = x$ and there are no fixed points.

At $\mu = \frac{1}{4}$ the parabola just touches the line $y = x$ at the point $(\frac{1}{2}, \frac{1}{2})$ and so

$$p = \frac{1}{2}$$

is a fixed point, with $Q'_\mu(p) = 2p = 1$.

For $\mu < \frac{1}{4}$ the points $p_\pm$ are fixed points, with $Q'_\mu(p_+) > 1$ so it is repelling, and $Q'_\mu(p_-) < 1$. We will have $Q'_\mu(p_-) > -1$ so long as $\mu > -\frac{3}{4}$, so on the range $-\frac{3}{4} < \mu < \frac{1}{4}$ we have two fixed points, one repelling and one attracting.

We will now discuss the general phenomenon. In order not to clutter up the notation, we assume that coordinates have been chosen so that $b = 0$ and $p = 0$. So we make the standing assumption that $p = 0$ is a fixed point at $\mu = 0$, i.e. that

$$F(0, 0) = 0.$$

Theorem 2.2.1. (Fold bifurcation). *Suppose that at the point* $(0, 0)$ *we have*

$$\text{(a)} \quad \frac{\partial F}{\partial x}(0,0) = 1, \qquad \text{(b)} \quad \frac{\partial^2 F}{\partial x^2}(0,0) > 0, \qquad \text{(c)} \quad \frac{\partial F}{\partial \mu}(0,0) > 0.$$

Then there are non-empty intervals $(\mu_1, 0)$ *and* $(0, \mu_2)$ *and* $\epsilon > 0$ *so that*

(i) *If* $\mu \in (\mu_1, 0)$ *then* F_μ *has two fixed points in* $(-\epsilon, \epsilon)$.
One is attracting and the other repelling.

(ii) *If* $\mu \in (0, \mu_2)$ *then* F_μ *has no fixed points in* $(-\epsilon, \epsilon)$.

Proof of the fold bifurcation theorem, step I.

The proofs in this section and the next will be applications of the implicit function theorem. For our current theorem, set

$$P(x, \mu) := F(x, \mu) - x.$$

Then by our standing hypothesis we have

$$P(0, 0) = 0$$

and condition (c) says that

$$\frac{\partial P}{\partial \mu}(0, 0) > 0.$$

The implicit function theorem gives a unique function $\mu(x)$ with $\mu(0) = 0$ and

$$P(x, \mu(x)) \equiv 0.$$

The formula for the derivative in the implicit function theorem gives

$$\mu'(x) = -\frac{\partial P/\partial x}{\partial P/\partial \mu}$$

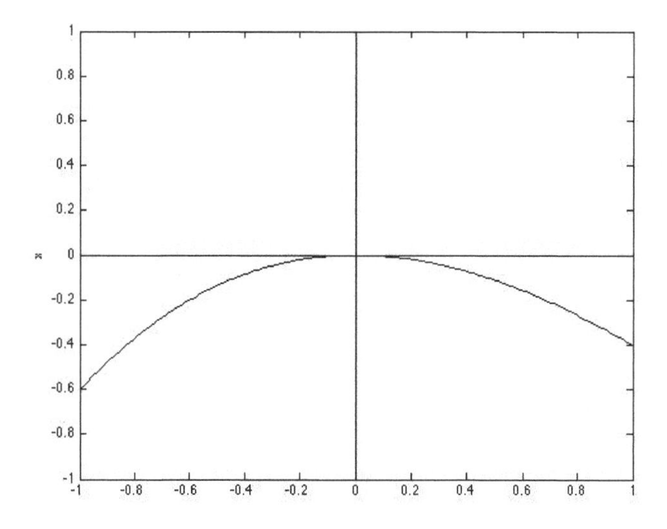

Figure 2.11: Graph of the function $x \mapsto \mu(x)$.

which vanishes at the origin by assumption (a). We then may compute the second derivative, μ'', via the chain rule; using the fact that $\mu'(0) = 0$ we obtain

$$\mu''(0) = -\frac{\partial^2 P/\partial x^2}{\partial P/\partial \mu}(0,0).$$

This is negative by assumptions (b) and (c).

Proof of the fold bifurcation theorem, step II.

In other words,

$$\mu'(0) = 0, \text{ and } \mu''(0) < 0$$

so $\mu(x)$ has a maximum at $x = 0$, and this maximum value is 0. In the (x, μ) plane, the graph of $\mu(x)$ looks locally approximately like a parabola in the lower half plane with its apex at the origin.

Proof of the fold bifurcation theorem, step III - rotate the picture.

If we rotate this picture clockwise by ninety degrees, this says that there are no points on this curve sitting over positive μ values, i.e. no fixed points for positive μ, and two fixed points for $\mu < 0$.

Proof of the fold bifurcation theorem, step IV.

We have established that for $|\mu|$ small there are two fixed points for $\mu < 0$ and no fixed points for $\mu > 0$. We must now show that one of these fixed points is

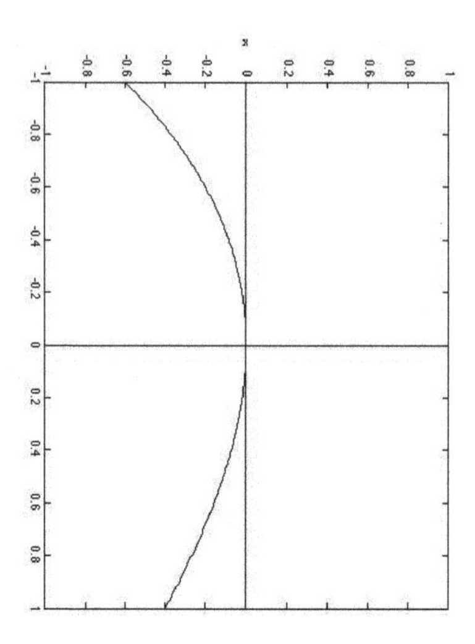

Figure 2.12: Rotating the preceding figure.

attracting and the other repelling.

For this, consider the function $\frac{\partial F}{\partial x}(x, \mu(x))$. The derivative of this function with respect to x is

$$\frac{\partial^2 F}{\partial x^2}(x, \mu(x)) + \frac{\partial^2 F}{\partial x \partial \mu}(x, \mu(x))\mu'(x).$$

Assumption (b) says that $\frac{\partial^2 F}{\partial x^2}(0,0) > 0$, and we know that $\mu'(0) = 0$. So the above expression is positive at 0.

Proof of the fold bifurcation theorem, step V, completion of the proof.

We know that $\frac{\partial F}{\partial x}(x, \mu(x))$ is an increasing function in a neighborhood of the origin while $\frac{\partial F}{\partial x}(0,0) = 1$. But this says that

$$F'_\mu(x) < 1$$

on the lower fixed point and

$$F'_\mu(x) > 1$$

at the upper fixed point, completing the proof of the theorem. $\square$

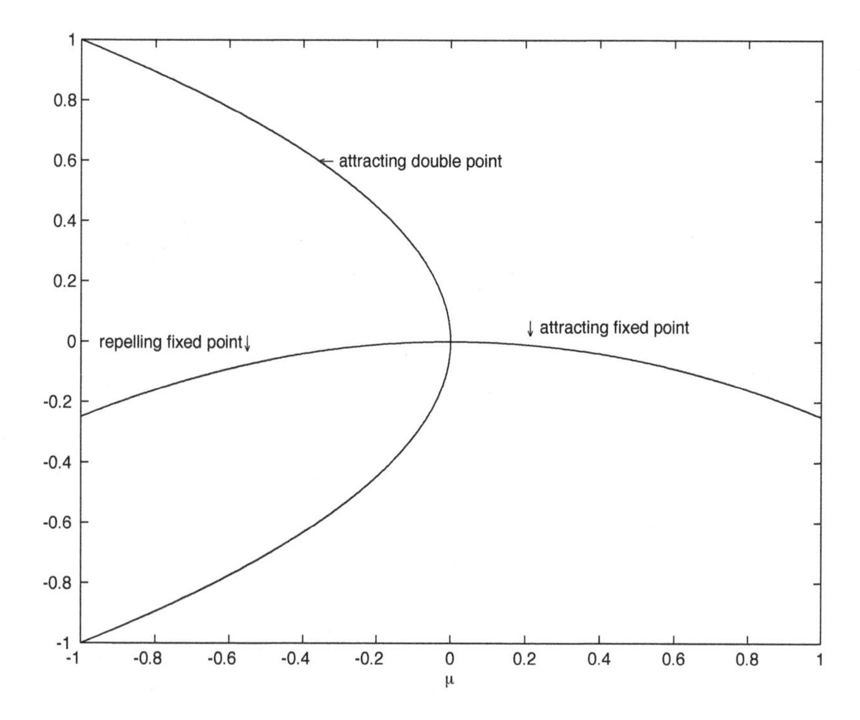

Figure 2.13: The period doubling bifurcation.

2.3 The period doubling bifurcation.

2.3.1 Description of the period doubling bifurcation.

The **fold** bifurcation arises when the parameter μ passes through a value where $F_\mu(x) = x$ and $F'_\mu(x) = 1$.

Under the appropriate hypotheses, the **period doubling** bifurcation describes what happens when μ passes through a bifurcation value b where $F_b(x) = x$ and $F'_\mu(x) = -1$.

On one side of b there is a single attractive fixed point. On the other side the attractive fixed point has become a repelling fixed point, and an attractive periodic point of period two has arisen.

An example.

Before stating the period doubling bifurcation theorem, we look at an example we have already considered: the first period doubling bifurcation in the logistic family, the bifurcation at $\mu = 3$. In Figure 2.14 we plot the function $L_\mu^{\circ 2}$ for the values $\mu = 2.9$ and $\mu = 3.3$. For $\mu = 2.9$ the curve crosses the diagonal at a single point, which is in fact a fixed point of L_μ and hence of $L_\mu^{\circ 2}$. This fixed

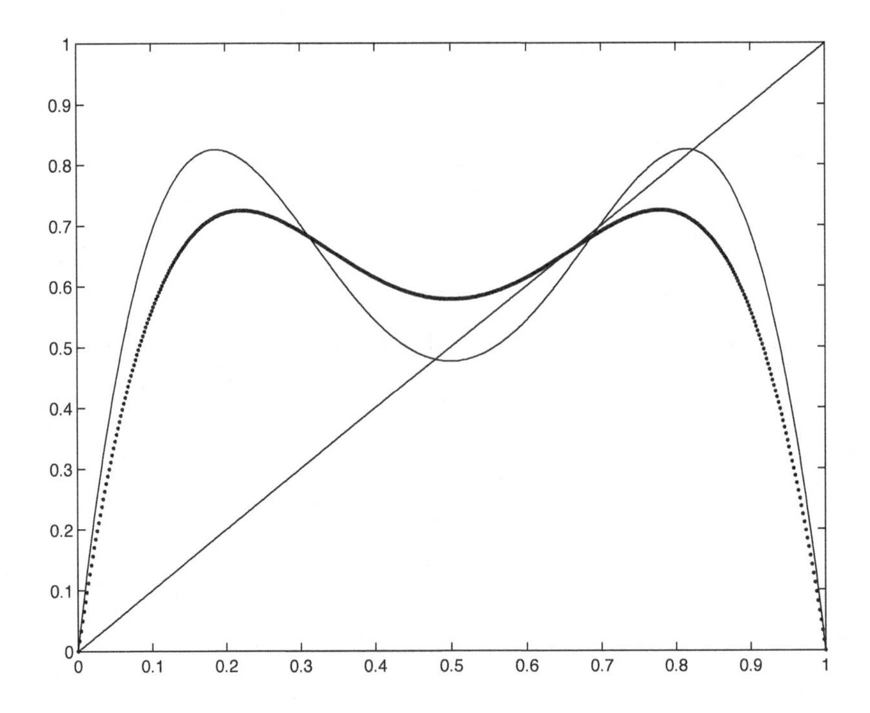

Figure 2.14: Plots of $L_\mu^{\circ 2}$ at $\mu = 2.9$ (dotted curve) and $\mu = 3.3$.

point is stable. For $\mu = 3.3$ there are three crossings. The non-zero fixed point of L_μ has derivative smaller than -1, and hence the corresponding fixed point of $L_\mu^{\circ 2}$ has derivative greater than one. The two other crossings correspond to the stable period two orbit.

2.3.2 Statement of the period doubling bifurcation theorem.

We now turn to the general theory: We are now assuming that $\mu = 0$ has 0 as a fixed point with $F_0'(0) = -1$. So the partial derivative of $F(x, \mu) - x$ with respect to x is -2 at $(0,0)$. In particular it does not vanish, so we can locally solve for x as a function of μ; there is (locally) a unique branch of fixed points, $x(\mu)$, passing through the origin.

Let $\lambda(\mu)$ denote the derivative of F_μ with respect to x at the fixed point, $x(\mu)$, i.e. define

$$\lambda(\mu) := \frac{\partial F}{\partial x}(x(\mu), \mu).$$

Recall that as notation, we are writing

$$F_\mu^{\circ 2} := F_\mu \circ F_\mu$$

and we define

$$F^{\circ 2}(x,\mu) := F_\mu^{\circ 2}(x).$$

Notice that

$$(F_\mu^{\circ 2})'(x) = F_\mu'(F_\mu(x))F_\mu'(x)$$

by the chain rule so

$$(F_0^{\circ 2})'(0) = (F_0'(0))^2 = 1.$$

Hence

$$(F_\mu^{\circ 2})''(x) = F_\mu''(F_\mu(x))F_\mu'(x)^2 + F_\mu'(F_\mu(x))F_\mu''(x) \tag{2.4}$$

which vanishes at $x = 0, \mu = 0$. In other words,

$$\frac{\partial^2 F^{\circ 2}}{\partial x^2}(0,0) = 0. \tag{2.5}$$

Let us absorb the import of this equation. One might think that if we set $G_\mu = F_\mu^{\circ 2}$, then $G_\mu'(0) = 1$, so all we need to do is apply the fold bifurcation theorem to G_μ. But (2.5) shows that the key condition (b) in the fold bifurcation theorem, namely:

$$\frac{\partial^2 F}{\partial x^2}(0,0) > 0,$$

is violated, and hence we must make some alternative hypotheses. The hypotheses that we will make will involve the second and the third partial derivatives of F, and also that $\lambda(\mu)$ really passes through -1, i.e. $\frac{d\lambda}{d\mu}(0) \neq 0$. To understand the hypothesis we will make involving the partial derivatives of F, let us differentiate (2.4) once more with respect to x to obtain

$$(F_\mu^{\circ 2})'''(x) =$$

$$F_\mu'''(F_\mu(x))F_\mu'(x)^3 + 2F_\mu''(F_\mu(x))F_\mu''(x)F_\mu'(x)$$
$$+F_\mu''(F_\mu(x))F_\mu'(x)F_\mu''(x) + F_\mu'(F_\mu(x))F_\mu'''(x).$$

At $(x,\mu) = (0,0)$ this simplifies to

$$-\left[2\frac{\partial^3 F}{\partial x^3}(0,0) + 3\left(\frac{\partial^2 F}{\partial x^2}(0,0)\right)^2\right]. \tag{2.6}$$

Theorem 2.3.1. [Period doubling bifurcation.] *Suppose that F is C^3, that*

$$\text{(d)} F_0'(0) = -1 \quad \text{(e)} \quad \frac{d\lambda}{d\mu}(0) > 0, \quad \textit{and}$$

$$\text{(f)} \quad 2\frac{\partial^3 F}{\partial x^3}(0,0) + 3\left(\frac{\partial^2 F}{\partial x^2}(0,0)\right)^2 > 0.$$

Then there are non-empty intervals $(\mu_1, 0)$ and $(0, \mu_2)$ and $\epsilon > 0$ so that

(i) *If $\mu \in (\mu_1, 0)$ then F_μ has one repelling fixed point and one attracting orbit of period two in $(-\epsilon,\epsilon)$*

(ii) *If $\mu \in (0, \mu_2)$ then $F_\mu^{\circ 2}$ has a single fixed point in $(-\epsilon,\epsilon)$ which is in fact an attracting fixed point of F_μ.*

The conclusions of the theorem are summarized in Figure 2.13.

The proof of the period doubling bifurcation theorem is considerably harder than the proof of the fold bifurcation theorem.

2.3.3 Proof of the period doubling bifurcation theorem.

Step I.

Let

$$H(x,\mu) := F^{\circ 2}(x,\mu) - x.$$

Then by the remarks before the statement of the theorem, H vanishes at the origin together with its first two partial derivatives with respect to x. Formula (2.6) (which used condition (d)) together with condition (f) gives

$$\frac{\partial^3 H}{\partial x^3}(0,0) < 0.$$

One of the zeros of $H(x,\mu) := F^{\circ 2}(x,\mu) - x$ at the origin corresponds to the fact that $(0,0)$ is a fixed point. Let us factor this out: Define $P(x,\mu)$ by

$$H(x,\mu) = (x - x(\mu))P(x,\mu). \tag{2.7}$$

Step II.

Then

$$\begin{aligned}
\frac{\partial H}{\partial x} &= P + (x - x(\mu))\frac{\partial P}{\partial x} \\
\frac{\partial^2 H}{\partial x^2} &= 2\frac{\partial P}{\partial x} + (x - x(\mu))\frac{\partial^2 P}{\partial x^2} \\
\frac{\partial^3 H}{\partial x^3} &= 3\frac{\partial^2 P}{\partial x^2} + (x - x(\mu))\frac{\partial^3 P}{\partial x^3}.
\end{aligned}$$

So P vanishes at the origin together with its first partial derivative with respect to x, while

$$\frac{\partial^3 H}{\partial x^3}(0,0) = 3\frac{\partial^2 P}{\partial x^2}(0,0)$$

so

$$\frac{\partial^2 P}{\partial x^2}(0,0) < 0. \tag{2.8}$$

Step III.

We claim that

$$\frac{\partial P}{\partial \mu}(0,0) < 0, \tag{2.9}$$

so that we can apply the implicit function theorem to $P(x,\mu) = 0$ to solve for μ as a function of x. This will allow us to determine the fixed points of $F_\mu^{\circ 2}$

which are *not* fixed points of F_μ, i.e. the points of period two. To prove (2.9) we compute $\frac{\partial H}{\partial x}$ both from its definition $H(x,\mu) = F^{\circ 2}(x,\mu) - x$ and from (2.7) to obtain:

$$\begin{aligned}\frac{\partial H}{\partial x} &= \frac{\partial F}{\partial x}(F(x,\mu),\mu)\frac{\partial F}{\partial x}(x,\mu) - 1 \\ &= P(x,\mu) + (x - x(\mu))\frac{\partial P}{\partial x}(x,\mu).\end{aligned}$$

Recall that $x(\mu)$ is the fixed point of F_μ and that $\lambda(\mu) = \frac{\partial F}{\partial x}(x(\mu),\mu)$. So substituting $x = x(\mu)$ into the preceding equation gives

$$\lambda(\mu)^2 - 1 = P(x,\mu).$$

Differentiating with respect to μ and setting $\mu = 0$ gives

$$\frac{\partial P}{\partial \mu}(0,0) = 2\lambda(0)\lambda'(0) = -2\lambda'(0)$$

which is < 0 by (e).

Step IV.

By the implicit function theorem, (2.9) implies that there is a C^2 function $\nu(x)$ defined near zero as the unique solution of $P(x,\nu(x)) \equiv 0$. Recall that P and its first derivative with respect to x vanish at $(0,0)$. We now repeat the arguments of the preceding section: We have

$$\nu'(x) = -\frac{\partial P/\partial x}{\partial P/\partial \mu}$$

so

$$\nu'(0) = 0$$

and

$$\nu''(0) = -\frac{\partial^2 P/\partial x^2}{\partial P/\partial \mu}(0,0) < 0$$

since this time both numerator and denominator are negative.

So the curve ν has the same form as in the proof of the fold bifurcation theorem. This establishes the existence of the (strictly) period two points for $\mu < 0$ and their absence for $\mu > 0$.

Step V.

We now turn to the question of the stability of the fixed points and the period two points. Condition (e):

$$\frac{d\lambda}{d\mu}(0) > 0,$$

together with the fact that $\lambda(0) = -1$ imply that $\lambda(\mu) < -1$ for $\mu < 0$ and $\lambda(\mu) > -1$ for $\mu > 0$ so the fixed point is repelling to the left and attracting to the right of the origin. As for the period two points, we wish to show that

$$\frac{\partial F^{\circ 2}}{\partial x}(x,\nu(x)) < 1$$

for $x < 0$.

Now (2.5) and $\nu'(0) = 0$ imply that 0 is a critical point for this function, and the value at this critical point is $\lambda(0)^2 = 1$. To complete the proof we must show that this critical point is a local maximum. So we must compute the second derivative at the origin.

Step VI, completion of the proof.

Calling this function ϕ we have

$$\begin{aligned}
\phi(x) &:= \frac{\partial F^{\circ 2}}{\partial x}(x,\nu(x)) \\
\phi'(x) &= \frac{\partial^2 F^{\circ 2}}{\partial x^2}(x,\nu(x)) + \frac{\partial^2 F^{\circ 2}}{\partial x \partial \mu}(x,\nu(x))\nu'(x) \\
\phi''(x) &= \frac{\partial^3 F^{\circ 2}}{\partial x^3}(x,\nu(x)) + 2\frac{\partial^3 F^{\circ 2}}{\partial x^2 \partial \mu}(x,\nu(x)\nu'(x) \\
&\quad + \frac{\partial^3 F^{\circ 2}}{\partial x \partial \mu^2}(x,\nu(x))(\nu'(x))^2 + \frac{\partial^2 F^{\circ 2}}{\partial x \partial \mu}(x,\nu(x))\nu''(x).
\end{aligned}$$

The middle two terms vanish at 0 since $\nu'(0) = 0$. The last term becomes

$$\frac{d\lambda}{d\mu}(0)\nu''(0) < 0$$

by condition (e) and the fact that $\nu''(0) < 0$. We have computed the the first term, i.e. the third partial derivative, in (2.6) using condition (d) and then (f) implies that this expression is negative. This completes the proof of the period doubling bifurcation theorem. □

Variants.

There are obvious variants on the theorem which involve changing signs in hypotheses (e) and or (f). Thus we may have an attractive fixed point merging with two repelling points of period two to produce a repelling fixed point, and/or the direction of the bifurcation may be reversed.

2.4 Newton's method and Feigenbaum's constant.

Although the bifurcation values b_n for the logistic family are hard to compute except by numerical experiment, the superattractive values can be found by applying Newton's method to find the solution, s_n, of the equation

$$L_\mu^{\circ 2^{n-1}}(\frac{1}{2}) = \frac{1}{2}, \quad L_\mu(x) = \mu x(1-x). \tag{2.10}$$

This is the equation for μ which says that $\frac{1}{2}$ is a point of period 2^{n-1} of L_μ. Of course we want to look for solutions for which $\frac{1}{2}$ does not have lower period.

So we set

$$P(\mu) = L_\mu^{\circ 2^{n-1}}(\frac{1}{2}) - \frac{1}{2}$$

and apply the Newton algorithm

$$\mu_{k+1} = \mathcal{N}(\mu_k), \quad \mathcal{N}(\mu) = \mu - \frac{P(\mu)}{P'(\mu)}.$$

with $'$ now denoting differentiation with respect to μ.

As a first step, must compute P and P'. For this we define the functions $x_k(\mu)$ recursively by

$$x_0 \equiv \frac{1}{2}, \quad x_1(\mu) = \mu\frac{1}{2}(1-\frac{1}{2}), \quad x_{k+1} = L_\mu(x_k),$$

so, we have

$$\begin{aligned} x'_{k+1} &= [\mu x_k(1-x_k))]' \\ &= x_k(1-x_k) + \mu x'_k(1-x_k) - \mu x_k x'_k \\ &= x_k(1-x_k) + \mu(1-2x_k)x'_k. \end{aligned}$$

Let

$$N = 2^{n-1}$$

so that

$$P(\mu) = x_N - \frac{1}{2}, \quad P'(\mu) = x'_N(\mu).$$

Thus, at each stage of the iteration in Newton's method we compute $P(\mu)$ and $P'(\mu)$ by running the iteration scheme

$$\begin{array}{lllll} x_{k+1} & = & \mu x_k(1-x_k) & x_0 & = & \frac{1}{2} \\ x'_{k+1} & = & x_k(1-x_k) + \mu(1-2x_k)x'_x & x'_0 & = & 0 \end{array}$$

for $k = 0, \ldots, N-1$. We substitute this into Newton's method, get the next value of μ, run the iteration to get the next value of $P(\mu)$ and $P'(\mu)$ etc.

Suppose we have found $s_1, s_2, \ldots., s_n$. What should we take as the initial value of μ? Define the numbers δ_n, $n \geq 2$ recursively by $\delta_2 = 4$ and

$$\delta_n = \frac{s_{n-1} - s_{n-2}}{s_n - s_{n-1}}, \quad n \geq 3. \tag{2.11}$$

We have already computed

$$s_1 = 2, \quad s_2 = 1 + \sqrt{5} = 3.23606797\ldots.$$

We take as our initial value in Newton's method for finding s_{n+1} the value

$$\mu_{n+1} = s_n + \frac{s_n - s_{n-1}}{\delta_n}.$$

The following facts are observed:

For each $n = 3, 4, \ldots, 15$, Newton's method converges very rapidly, with no changes in the first nineteen digits after six applications of Newton's method for finding s_3, after only one application of Newton's method for s_4 and s_5, and at most four applications of Newton's method for the computation of each of the remaining values.

Suppose we stop our calculations for each s_n when there is no further change in the first 19 digits, and take the computed values as our s_n. These values are strictly increasing. In particular this implies that the s_n we have computed do not yield $\frac{1}{2}$ as a point of lower period.

The s_n approach a limiting value, 3.569945671205296863.

The δ_n approach a limiting value,

$$\delta = 4.6692016148.$$

This value is known as **Feigenbaum's constant**. While the limiting value of the s_n is particular to the logistic family, δ is "universal" in the sense that it applies to a whole class of one dimensional iteration families. We shall go into this point in the next section, where we will see that this is a "renormalization group" phenomenon.

2.5 Feigenbaum renormalization.

We have already remarked that the rate of convergence to the limiting value of the superstable points in the period doubling bifurcation, Feigenbaum's constant, is universal, i.e. not restricted to the logistic family. That is, if we let

$$\delta = 4.6692....$$

denote Feigenbaum's constant, then the superstable values s_r in the period doubling scenario satisfy

$$s_r = s_\infty - B\delta^{-r} + o(\delta^{-r})$$

where s_∞ and B depend on the specifics of the family, but δ applies to a large class of such families.

There is another "universal" parameter in the story. Suppose that our family f_μ consists of maps with a single maximum, X_m, so that X_m must be one of the points on any superstable periodic orbit. (In the case of the logistic family $X_m = \frac{1}{2}$.) Let d_r denote the difference between X_m an the next nearest point on the superstable 2^r orbit; more precisely, define

$$d_r = f_{s_r}^{2^{r-1}}(X_m) - X_m.$$

$$-\alpha z_{2j} = z_j$$

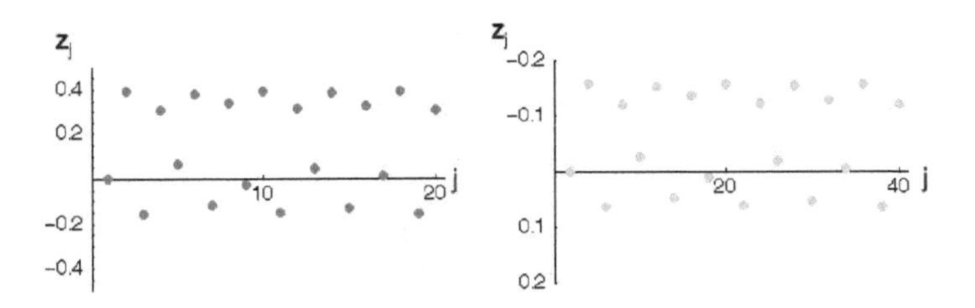

Figure 2.15: From a powerpoint presentation by Prof. Coppersmith, reproduced with her permission.

Then $\quad d_r \sim D(-\alpha)^r \quad$ where

$$\alpha \doteq 2.5029...$$

is again universal. This would appear to be a scale parameter (in x) associated with the period doubling scenario.

To understand this scale parameter, examine the central portion of Figure 2.5 , and observe that the graph of $L_\mu^{\circ 2}$ looks like an (inverted and) rescaled version of L_μ, especially if we allow a change in the parameter μ.

Before going into the rescaling operator on functions, I would like to give an elementary formulation of what is going on, following a beautiful paper by S.N. Coppersmith *A simpler derivation of Feigenbaum's renormalization group equation for the period-doubling bifurcation sequence* which appeared in the American Journal of Physics, Vol **67** (1999) 53. Also see her powerpoint presentation available on the web.

Take $\mu = 3.569946$ and plot the values $L_\mu^j(.5) - .5$ against j. Then plot every other value with the ordinate upside down and rescaled by a factor of 2.502 907 9. The graphs look the same: In the left hand of the figure, every j is plotted; in the right figure, every other j is plotted with the ordinate upside down.

We can check this numerically by computing the vector y (say of length 21) with $y(1) = .5$ and $y(n+1) = L_\mu(y(n))$, then the vector z with $z(i) = y(i) - .5$ and then comparing the first 11 entries of z with the vector k obtained by taking every other entry of z and multiplying by -2.502. The results are:

0	0
0.3925	0.3939
−0.1574	−0.1566
0.3040	0.2976
0.0626	0.0626
0.3785	0.3783
−0.1190	−0.1189
0.3420	0.3380
−0.0250	−0.0250
0.3903	0.3914
−0.1512	−0.1505

The existence of the scaling (together with some argumentation) determines the scale parameter as follows: We presume to have

$$-\alpha z_{2j} = z_j$$

which, replacing j by $j+1$ gives $-\alpha z_{2j+2} = z_{j+1}$. Write $z_{j+1} = g(z_j)$. The second equation gives $-\alpha g(g(z_{2j}) = g(z_j)$ and we can substitute $z_{2j} = -z_j/\alpha$ from the first equation to get

$$-\alpha g(g(-z_j/\alpha)) = g(z_j).$$

If we expect this to hold not just for z_j but for all values of z we get the functional equation:

$$-\alpha g(g(-z/\alpha)) = g(z).$$

If we assume that g has a power series expansion, and we compute up to terms of second order in z, we get an approximate value for α.

The rescaling is centered at the maximum, so in order to avoid notational complexity, let us shift this maximum (for the logistic family) to the origin by replacing x by $y = x - \frac{1}{2}$. In the new coordinates the logistic map is given by

$$y \mapsto L_\mu(y + \frac{1}{2}) - \frac{1}{2} = \mu(\frac{1}{4} - y^2) - \frac{1}{2}.$$

Let $\mathcal{R}$ denote the operator on functions given by

$$\mathcal{R}(h)(y) := -\alpha h(h(y/(-\alpha))). \tag{2.12}$$

In other words, $\mathcal{R}$ sends a map h into its iterate $h \circ h$ followed by a rescaling.

We are going to not only apply the operator $\mathcal{R}$, but also shift the parameter μ in the maps

$$h_\mu(y) = \mu(\frac{1}{2} - y^2) - \frac{1}{2}$$

from one supercritical value to the next. So for each $k = 0, 1, 2, \ldots$ we set

$$g_{k0} := h_{s_k}$$

and then define

$$\begin{aligned} g_{k,1} &= \mathcal{R}g_{k+1,0} \\ g_{k,2} &= \mathcal{R}g_{k+1,1} \\ g_{k,3} &= \mathcal{R}g_{k+2,1} \\ \vdots & \quad \vdots \end{aligned}$$

It is observed (numerically) that for each k the functions $g_{k,r}$ appear to be approaching a limit, g_k i.e.

$$g_{k,r} \to g_k.$$

So

$$g_k(y) = \lim(-\alpha)^r g_{s_{k+r}}^{2^r}(y/(-\alpha)^r).$$

Hence

$$\mathcal{R}g_k = \lim(-\alpha)^{r+1} 2^{r+1} g_{s_{k+r}}(y/(-\alpha)^{r+1}) = g_{k-1}.$$

It is also observed that these limit functions g_k themselves are approaching a limit:

$$g_k \to g.$$

Since $\mathcal{R}g_k = g_{k-1}$ we conclude that

$$\mathcal{R}g = g,$$

i.e. g is a fixed point for the Feigenbaum renormalization operator $\mathcal{R}$.

Notice that rescaling commutes with $\mathcal{R}$: If S denotes the operator $(Sf)(y) = cf(y/c)$ then

$$\mathcal{R}(Sf)(y) = -\alpha(c(f(cf(y/(c\alpha))/c) = S(\mathcal{R})f(y).$$

So if g is a fixed point, so is Sg. We may thus fix the scale in g by requiring that

$$g(0) = 1.$$

The hope was then that there would be a unique function g (within an appropriate class of functions) satisfying

$$\mathcal{R}g = g, \quad g(0) = 1,$$

or, spelling this out,

$$g(y) = -\alpha g^{\circ 2}(-y/\alpha), \quad g(0) = 1. \tag{2.13}$$

Notice that if we knew the function g, then setting $y = 0$ in (2.13) gives

$$1 = -\alpha g(1)$$

or

$$\alpha = -1/g(1).$$

In other words, assuming that we were able to establish all these facts and also knew the function g, then the universal rescaling factor α would be determined by g itself. Feigenbaum assumed that g has a power series expansion in x^2 took the first seven terms in this expansion and substituted in (2.13). He obtained a collection of algebraic equations which he solved and then derived α close to the observed "experimental" value. Indeed, if we truncate (2.13) we will get a collection of algebraic equations. But these equations are not recursive, so that at each stage of truncation modification is made in all the coefficients, and also the nature of the solutions of these equations is not transparent.

So theoretically, if we could establish the existence of a unique solution to (2.13) within a given class of functions the value of α is determined. But the numerical evaluation of α is achieved by the renormalization property itself, rather than from $g(1)$ which is not known explicitly.

The other universal constant associated with the period doubling scenario, the constant δ was also conjectured by Feigenbaum to be associated to the fixed point g of the renormalization operator; this time with the linearized map J, i.e. the derivative of the renormalization operator at its fixed point.

Later on we will see that in finite dimensions, if the derivative J of a non-linear transformation R at a fixed point has k eigenvalues > 1 in absolute value, and the rest < 1 in absolute value, then there exists a k-dimensional R invariant surface tangent at the fixed point to the subspace corresponding to the k eigenvalues whose absolute value is > 1. On this invariant manifold, the map R is expanding. Feigenbaum conjectured that for the operator $\mathcal{R}$ (acting on the appropriate infinite dimensional space of functions) there is a one dimensional "expanding" submanifold, and that δ is the single eigenvalue of J with absolute value greater than 1.

In the course of the past thirty five years, these conjectures of Feigenbaum have been verified using high powered techniques from complex analysis, thanks to the combined effort of such mathematicians as Douady, Hubbard, Sullivan, McMullen, and others. See, for example, [McMullen].

Chapter 3

Sarkovsky's theorem, Singer's theorem, intermittency.

The logistic map L_μ develops a period three orbit as μ increases above $1+\sqrt{8}$, and this orbit is initially stable. According to a theorem of Sarkovsky, if a continuous map of a compact interval on the real line has an orbit of period three, then it has orbits of all periods. Nevertheless, we do not see these other periodic orbits near $1+\sqrt{8}$. The reason is that they are all unstable. In fact, a theorem of Singer implies that L_μ can have at most one stable periodic orbit. In this chapter we explain these ideas, and also describe what happens to the period there orbits when we decrease μ from slightly above the critical value $1+\sqrt{8}$ to slightly below it.

Throughout this chapter, f will denote a continuous function on the reals whose domain of definition is assumed to include the given intervals in the various statements.

3.1 Period 3 implies all periods.

Lemma 3.1.1. *If $I = [a, b]$ is a compact interval and $I \subset f(I)$ then f has a fixed point in I.*

Proof. For some $c, d \in I$ we have $f(c) = a, f(d) = b$. So $f(c) \leq c, f(d) \geq d$. So $f(x) - x$ changes sign from c to d hence has a zero in between. □

Lemma 3.1.2. *If J and $K = [a, b]$ are compact intervals with $K \subset f(J)$ then there is a compact subinterval $L \subset J$ such that $f(L) = K$.*

Proof. Let c be the greatest point in J with $f(c) = a$. If $f(x) = b$ for some $x > c, x \in J$ let d be the least. Then we may take $L = [c, d]$. If not, $f(x) = b$ for some $x < c, x \in J$. Let c' be the largest. Let d' be the the smallest x satisfying $x > c'$ with $f(x) = a$. Notice that $d' \leq c$. We then take $L = [c', d']$. □

Notation. If I is a closed interval with end points a and b we write

$$I =< a, b >$$

when we do not want to specify which of the two end points is the larger.

Theorem 3.1.1. [Sarkovsky.] *Period three implies all periods.*

Proof. Suppose that f has a 3-cycle

$$a \mapsto b \mapsto c \mapsto a \mapsto \cdots.$$

Let a denote the leftmost of the three, and let us assume that

$$a < b < c.$$

Reversing left and right (i.e. changing direction on the real line) and cycling through the points makes this assumption harmless. Indeed, if $a < c < b$ then we have the cycle, $b \mapsto c \mapsto a \mapsto b$ with $b > c > a$. So we assume that $a < b < c$. Let

$$I_0 = [a, b], \quad I_1 = [b, c]$$

so we have

$$f(I_0) \supset I_1, \quad f(I_1) \supset I_0 \cup I_1.$$

By Lemma 2 the fact that $f(I_1) \supset I_1$ implies that there is a compact interval $A_1 \subset I_1$ with $f(A_1) = I_1$. Since $f(A_1) = I_1 \supset A_1$ there is a compact subinterval $A_2 \subset A_1$ with $f(A_2) = A_1$. So

$$A_2 \subset A_1 \subset I, \quad f^{\circ 2}(A_2) = I_1.$$

By induction proceed to find compact intervals with

$$A_{n-2} \subset A_{n-3} \subset \cdots \subset A_2 \subset A_1 \subset I_1$$

with

$$f^{\circ(n-2)}(A_{n-2}) = I_1.$$

Since $f(I_0) \supset I_1 \supset A_{n-2}$ there is an interval $A_{n-1} \subset I_0$ with $f(A_{n-1}) = A_{n-2}$. Finally, since $f(I_1) \supset I_0$ there is a compact interval $A_n \subset I_1$ with $f(A_n) = A_{n-1}$. So we have

$$A_n \to A_{n-1} \to \cdots \to A_1 \to I_1$$

where each interval maps onto the next and $A_n \subset I_1$. By Lemma 3.1.1, f^n has a fixed point, x, in A_n. But $f(x)$ lies in I_0 and all the higher iterates up to n lie in I_1 so the period can not be smaller than n. So there is a periodic point of any period $n \geq 3$.

Since $f(I_1) \supset I_1$ there is a fixed point in I_1, and since $f(I_0) \supset I_1, f(I_1) \supset I_0$ there is a point of period two in I_0 which is not a fixed point of f. □

A more refined analysis which we will omit shows that period 5 implies the existence of all periods greater than 5 and period 2 and 4 (but not period 3). In general any odd period implies the existence of periods of all higher order (and all smaller even order).

3.1.1 The Sarkovsky ordering

In fact, Sarkovsky introduced the following ordering on the positive integers, and proved that the existence of a periodic point of f with (minimal) period equal to any integer implies the existence of periodic points with periods equal to any integer later in the ordering. Here is his ordering:

$$3, 5, 7, 9, 11, \ldots$$
$$2\cdot 3,\ 2\cdot 5,\ 2\cdot 7\ldots$$
$$2^2\cdot 3,\ 2^2\cdot 5,\ 2^2\cdot 7\ldots$$
$$\vdots$$
$$\ldots, 2^n \ldots, 2^3, 2^2, 2, 1.$$

We will not prove Sarkovski's theorem here, but refer to the literature. For example to [Devaney(1989)], page 63.

Notice that Sarkovky's theorem implies that if there are only finitely many periodic points, their periods must be powers of two. This was illustrated in the period doubling bifurcation at the beginning of the logistic family.

3.1.2 Periodic points of period three for the logistic family.

It is easy to graph the third iterate of the logistic map to see that it crosses the diagonal for $\mu > 1 + \sqrt(8)$ (in addition to the crossing corresponding to the fixed point). In fact, one can prove that that at $\mu = 1 + \sqrt(8)$ the graph of $L_\mu^{\circ 3}$ just touches the diagonal and strictly crosses it for $\mu > 1 + \sqrt(8) = 3.8284\ldots$. Hence in this range there are periodic points of all periods. Here are plots of L_μ for $\mu = 3.7, 3.81, 3.83, 3.84$: For $\mu = 1 + \sqrt{8} + .002$ the eight roots of $P(x) - x$ where $P := L_\mu^{\circ 3}$ and the values of P' at these roots are given by

x_j = roots of $P(x) - x$	$P'(x)$
0	56.20068544683054
0.95756178779471	0.24278522730018
0.95516891475013	1.73457935568109
0.73893250871724	−6.13277919589328
0.52522791460709	1.73457935766594
0.50342728916956	0.24278522531345
0.16402371217410	1.73457935778151
0.15565787278717	0.24278522521922

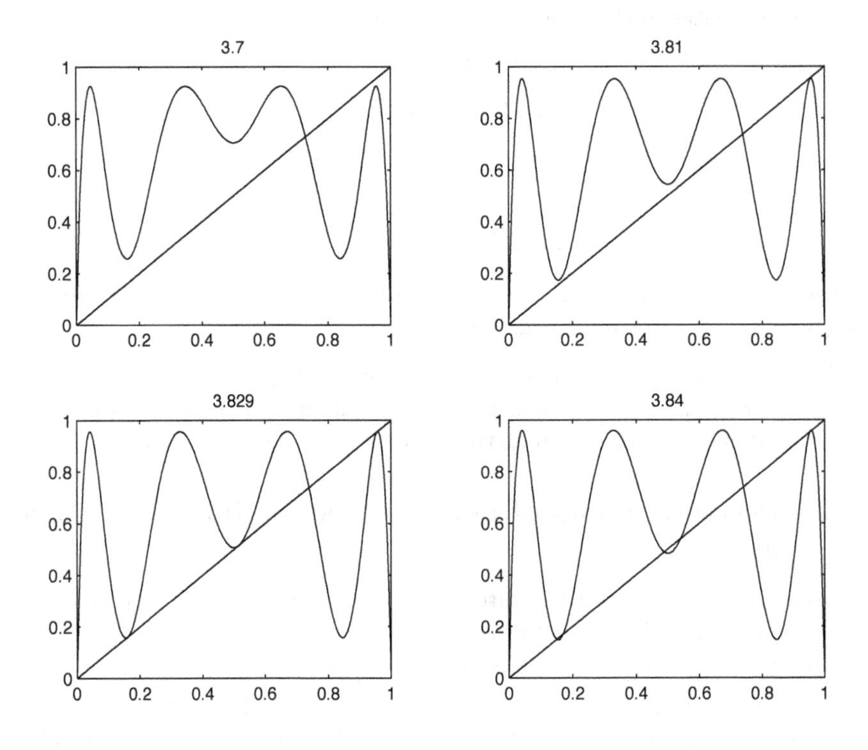

Figure 3.1: Graphs of $L_{\mu}^{\circ 3}$ for values of μ near $1 + \sqrt{8}$.

We see that there is a attractive period three orbit consisting of the points $0.1556\ldots, .5034\ldots, .9575\ldots$.. The fixed points 0 and .7389... are unstable and there is an unstable period three orbit.

There are, in fact, no other attractive periodic orbits of any period. We discuss the theoretical reason for this in the next section. It turns out, as a consequence of a theorem of David Singer (1978), which we discuss in the next section, there can no more than one attractive periodic orbit of L_μ for any value of μ!

3.2 Singer's theorem.

3.2.1 The Schwarzian derivative and some of its properties.

Let f be a function defined on some interval J and have three continuous continuous derivatives there. Define

$$(Sf)(x) = \frac{f'''(x)}{f'(x)} - \frac{3}{2}\left(\frac{f''(x)}{f'(x)}\right)^2. \tag{3.1}$$

$S(f)(x)$ is called the **Schwarzian derivative** of f at x whenever it exists as a number or as $\pm\infty$.

The Schwarzian derivative of the composite of two functions.

An important property of the Schwartzian derivative relates to the composite of two functions. Namely

$$S(f(g)) = S(f)(g)[g']^2 + S(g). \tag{3.2}$$

As a consequence, if both $S(f)$ and $S(g)$ are negative, so is $S(f(g))$. Here is the messy but straightforward proof of this fact: Using the formula for the derivative of $f \circ g$ we have

$$\begin{aligned}
(f\circ g)'(x) &= [f'(g(x)])g'(x),\\
(f\circ g)''(x) &= [f''(g(x))]g'(x)^2 + [f'(g(x)]g''(x),\\
(f\circ g)'''(x) &= [f'''(g(x))]g'(x)^3 + 3[f''(g(x))][g'(x)][g''(x)] + [f'(g(x))]g'''(x).
\end{aligned}$$

so from the definition (3.1) we have $S(f\circ g)(x) =$

$$\frac{[f'''(g(x))]g'(x)^3 + 3[f''(g(x))][g'(x)][g''(x)] + [f'(g(x))]g'''(x)}{f'(g(x)])g'(x)}$$

$$-\frac{3}{2}\left(\frac{[f''(g(x))]g'(x)^2 + [f'(g(x)]g''(x)}{[f'(g(x)])g'(x)}\right)^2$$

Collecting terms gives (3.2). □

A consequence of (3.2) is

Proposition 3.2.1. *If $S(f) < 0$ and $S(g) < 0$ then $S(f \circ g) < 0$.*

It follows by induction that if $S(f) < 0$ at all points then so is $S(f^{\circ n})$ for any positive integer n. So if we want to prove something about periodic cycles of a function f with $S(f) < 0$, then it will frequently be enough to prove a theorem about fixed points (of $f^{\circ n}$).

The idea of the statement and proof of Singer's theorem is that if f satisfies $Sf < 0$ (as we will assume until the end of this section) then all but 2 of the attracting cycles of f contain a critical point in their basin of attraction. The goal of the next few lemmas is to establish the existence of critical points (points where $f'(x) = 0$) from the existence of fixed points.

In what follows, we are assuming that $S(g) < 0$. In particular, since $S(\ell) \equiv 0$ for any linear function ℓ, a function g satisfying $S(g) < 0$ can not be identically equal to x on any open interval. We follow the treatment in [Gulick].

Critical points and fixed points for functions with negative Schwarzian derivative.

Lemma 3.2.1. *Let g be such that $S(g) < 0$. If g' has a relative minimum at x^* then $g'(x^*) < 0$. If g has a relative maximum at x^* then $g'(x^*) > 0$.*

Proof. If x^* is an extremal value of g', then $g''(x^*) = 0$ so

$$S(g)(x^*) = \frac{g'''(x^*)}{g'(x^*)}.$$

At a relative minimum of g', the numerator must be ≥ 0 so $S(g) < 0$ implies that the numerator is > 0 and the denominator is < 0. The reverse at a relative maximum. □

We continue to assume that $S(g) < 0$.

Lemma 3.2.2. *Let $a < b < c$ be fixed points of g. If $g'(b) \leq 1$ then g has a critical point in (a, c).*

Proof. Notice that $g(x) - x$ vanishes at a and b, so the mean value theorem implies that there is an r with $a < r < b$ with $g'(r) = 1$. Similarly, there is an s with $b < s < c$ and $g'(s) = 1$. If $g'(b) \leq 1$ then g' has a relative minimum in (s, t) which must be negative by Lemma 3.2.1. But since $g'(s) = g'(t) = 1 > 0$, there must be a point in (s, t) where $g' = 0$. □

Lemma 3.2.3. *Suppose that $a < b < c < d$ are fixed points of g. Then g has a critical point on (a, d).*

Proof. If $g'(b) \leq 1$ then g has a fixed point on (a, c) by Lemma 3.2.2, and if $g'(c) \leq 1$ then g has a fixed point on (b, d) by the same lemma. So we need to prove the lemma only in the case where $g'(b) > 1$ and $g'(c) > 1$. The fact that

$g'(b) = b$ and $g'(b) > 1$ implies that for $r > b$ and sufficiently close to b we have $g(r) > r$. Similarly, for $t < c$ and sufficiently close to c we have $g(t) < t$. By the mean value theorem, there is a point s with $r < s < t$ such that $g'(s) < 1$. Since $g'(b) > 1$ and $g'(c) > 1$, the function g' must have a relative minimum at some point $y \in (b, c)$. By Lemma 3.2.1, $g'(y) < 0$. Since $g'(y) < 0$ and $g'(c) > 0$ there must be some $z \in (y, c)$ such that $g'(s) = 0$. □

To summarize: If $S(g) < 0$ then four fixed points on an interval I implies the existence of a critical point in the interior of I.

Lemma 3.2.4. *f f has a finitely many critical points, then so does $f^{\circ m}$ for any m.*

Proof. If x is a critical point of $f^{\circ 2}$ then by the chain rule either x or $f(x)$ is a critical point of f. There can be only finitely many points $x, y, \cdots$ with $f(x) = f(y) = \ldots$ since $f(x) = f(y)$ imples that there is a z between x and y with $f'(z) = 0$. So $f^{\circ 2}$ has only finitely many critical points. Now proceed by induction. □

Lemma 3.2.5. *If f has finitely many critical points and $Sf < 0$ then $f^{\circ m}$ has finitely many fixed points for any m.*

Proof. $f^{\circ m}$ has finitely many critical points by Lemma 6. List the fixed points $a_1 < a_2 < \cdots$ of $f^{\circ m}$ in increasing order. Between a_1 and a_4 there is a critical point by Lemma 3. Between a_4 and a_7 there is another critical point by the same lemma. Etc. So there can not be infinitely many fixed points.

□

3.2.2 Proof and statement of Singer's theorem.

Let f satisfy $S(f) < 0$ where f is defined on a closed interval $J = [A, B]$, $-\infty \leq A$, $B \leq +\infty$, and suppose that $f(J) \subset J$.

The following argument will show that any attracting fixed point p of $f^{\circ m}$ must contain a critical point in its basin of attraction, except possibly for an attracting fixed point in an interval of the form $[A, a)$ or an interval of the form $(b, B]$.

So let p be an attracting fixed point of $g := f^{\circ m}$. Let (L, R) be the largest open interval containing p all of whose points are attracted to p by g. So there are three possibilities:

- $g(L) = L$ and $g(R) = R$,
- $g(L) = R$ and $g(R) = L$, or
- $g(L) = g(R)$.

In the first case, L, p and R are fixed points of g, so there is a critical point z of g in (L, R) by Lemma 2. So one of the points $z, f(z), \ldots f^{\circ m-1}(z)$ is a critical point of f, and one of them is attracted to p by iterates of g. So we have found a critical point of f which is attracted to the orbit of p under f.

In the second case, $g^{\circ 2}(L) = L$ and $g^{\circ 2}(R) = R$ and we can apply the preceding argument.

In the third case, since $L \neq R$ and $g(L) = g(R)$ there must be a critical point of g in $(L.R)$ by the mean value theorem. So we have proved:

Theorem 3.2.1. [David Singer.] *If p is an attractive fixed point of some iterate of f such that the largest open interval (L, R) in its basin of attraction satisfies $A < L$ and $R < B$ then there is a critical point of f which is attracted to the corresponding orbit of f. Since a point can be attracted to at most one orbit, it follows that if there are n critical points, there can be at most $n + 2$ attractive cycles, the two possible additional attractive periodic orbits have basins of attraction containing intervals of the form $[A, a)$ or $(b, B]$.*

3.2.3 Application to the logistic family.

For the logistic function $f = f_\mu = \mu x(1 - x)$ we have $f''' \equiv 0$, so $S(f) < 0$ for $x \neq \frac{1}{2}$ while $\lim_{x \to \frac{1}{2}} S(f)(x) = -\infty$ which also counts as $S(f)(x)) < 0$. Also the only critical point is at $x = \frac{1}{2}$.

We know that for $\mu \leq 1$ the point 0 is the only fixed point and attracts the whole interval. 0 becomes repelling for $\mu > 1$ and so there is no periodic cycle which contains $[0, a)$ in its basin of attraction. Since $f(1) = 0$, we also know that there is no periodic cycle which contains an interval of the form $(b, 1]$ in its basin of attraction. So the two possible additional cases in Singer's theorem are excluded, and L_μ has at most one attractive cycle for any μ.

3.3 Intermittency.

In this section we describe what happens to the period three orbits of L_μ as we decrease μ from slightly above the critical value $1 + \sqrt{8}$ to slightly below it. For $\mu = 1 + \sqrt{8} + .002$ recall that there is a stable period three orbit consisting of the points

$$0.1556\ldots, .5034\ldots, .9575\ldots.$$

If we choose our initial value of x close to $.5034\ldots$ and plot the successive 199 iterates of L_μ applied to x we obtain the upper graph in Figure 3.2. The lower graph gives $x(j + 3) - x(j)$ for $j = 1$ to 197. We will now decrease the parameter μ by .002 so that $\mu = 1 + \sqrt{8}$ is the parameter giving the onset of period three. For this value of the parameter, the graph of $P = L_\mu^{\circ 3}$ just touches the line $y = x$ at the three double roots of $P(x) - x$ which are at $0.1599288\ldots$, $0.514355\ldots.$, $0.9563180\ldots.$ (Of course, the eighth degree

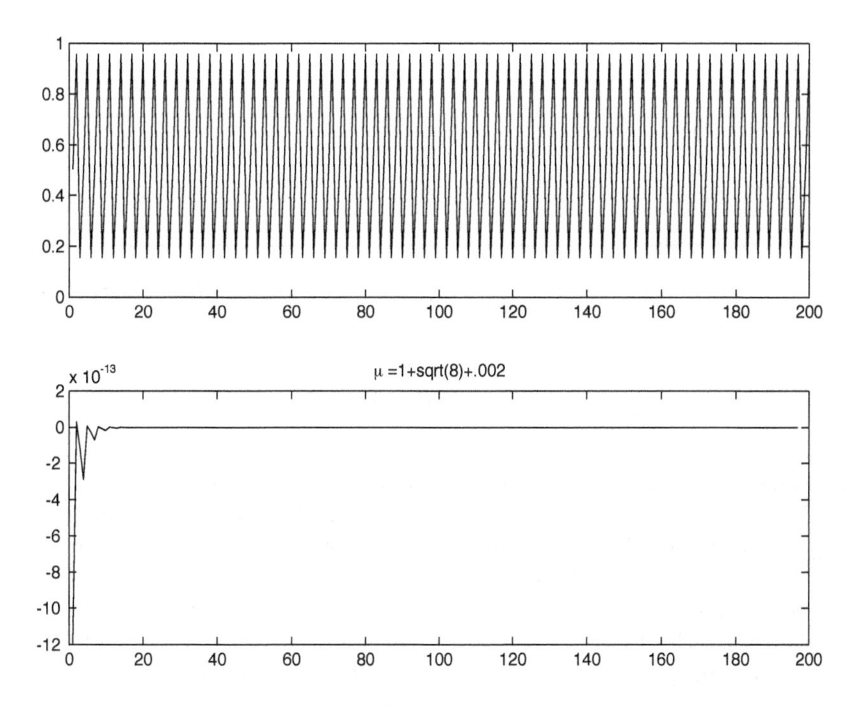

Figure 3.2: $\mu = 1 + \sqrt{8} + .002$.

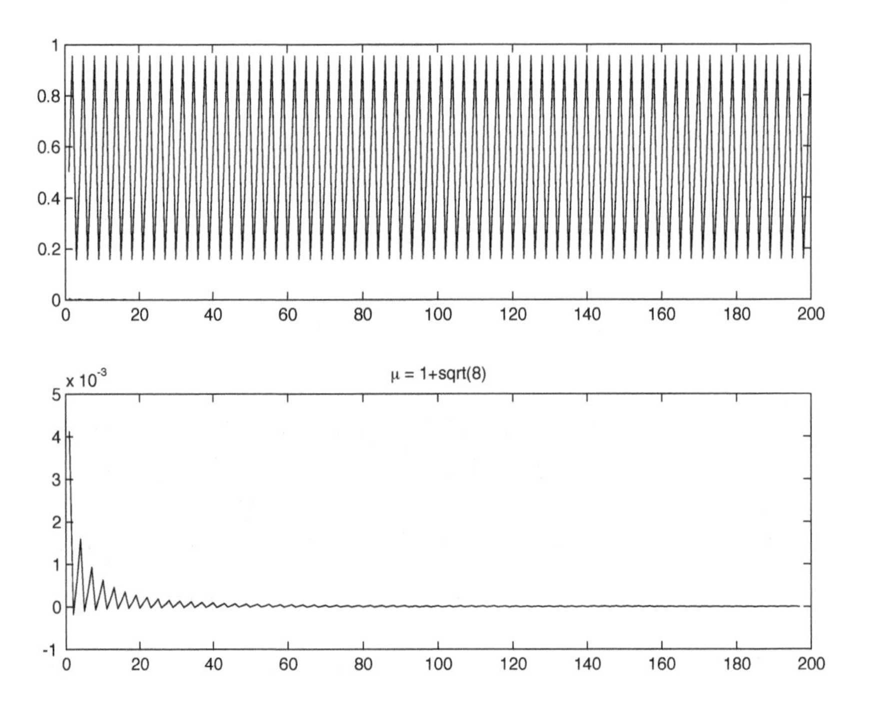

Figure 3.3: $\mu = 1 + \sqrt{8}$.

polynomial $P(x) - x$ has two additional roots which correspond to the two (unstable) fixed points of L_μ; these are not of interest to us.) Since the graph of P is tangent to the diagonal at the double roots, $P'(x) = 1$ at these points, so the period three orbit is not strictly speaking stable. But using the same initial seed as above, we do get slow convergence to the period three orbit, as is indicated by the Figure 3.3: Most interesting is what happens just before the onset of the period three cycle. Then $P(x) - x$ has only two real roots corresponding to the fixed points of L_μ. The remaining six roots are complex. Nevertheless, if μ is close to $1 + \sqrt{8}$ the effects of these complex roots can be felt. In Figure 3.4 we have taken $\mu = 1 + \sqrt{8} - .002$ and used the same initial seed $x = .5034$ and again plotted the successive 199 iterates of L_μ applied to x in the upper graph. Notice that there are portions of this graph where the behavior is almost as if we were at a point of period three, followed by some random looking behavior, then almost period three again and so on. This is seen more clearly in the bottom graph of $x(j+3) - x(j)$. Thus the bottom graph indicates that the deviation from period three is small on the j intervals $j = [1, 20]$, $[41, 65]$, $[96, 108]$, $[119, 124]$, $[148, 159]$, $[190, ?]$. This phenomenon is known as **intermittency**. We can understand how it works by graphically iterating $P = L_\mu^{\circ 3}$. As we pass close to a minimum of P lying just above the diagonal, or to a maximum of P lying just below the diagonal it will take many iterative steps to move away from this region - known as a **bottleneck**. Each

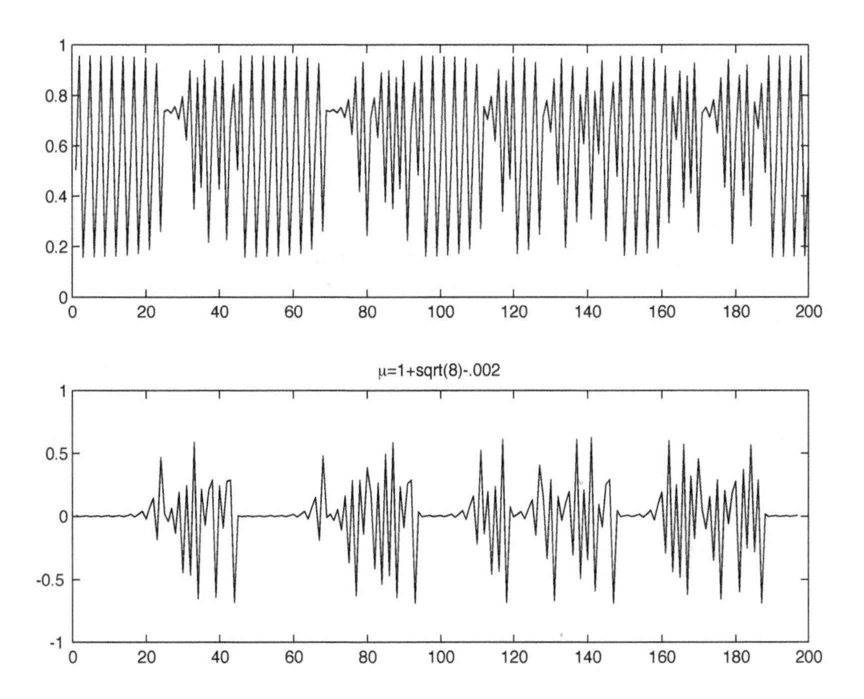

Figure 3.4: $\mu = 1 + \sqrt{8} - .002$.

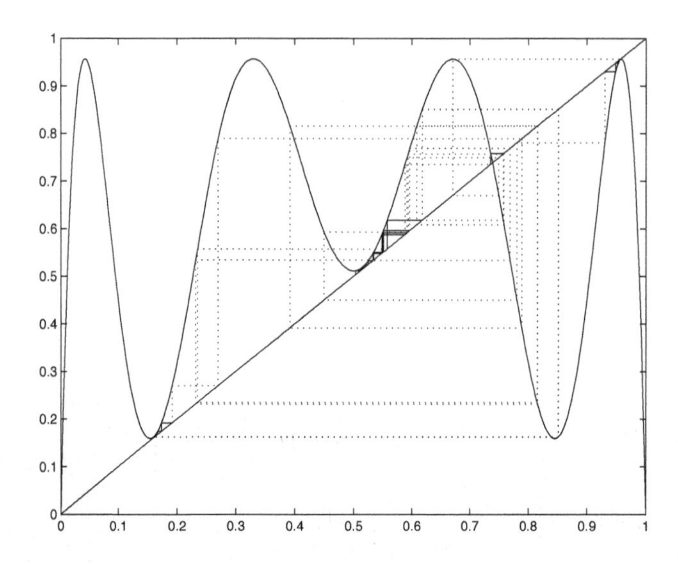

Figure 3.5: Graphical iteration of $P = L_\mu^{\circ 3}$ with $\mu = 1 + \sqrt{8} - .002$ and initial point .5034. The solid lines are iteration steps of size less than .07 representing bottleneck steps. The dotted lines are the longer steps.

such step corresponds to an almost period three cycle. After moving away from these bottlenecks, the steps will be large, eventually hitting a bottleneck once again. See the Figures 3.5 and 3.6.

The solid lines are iteration steps of size less than .07 representing bottleneck steps. The dotted lines are the longer steps.

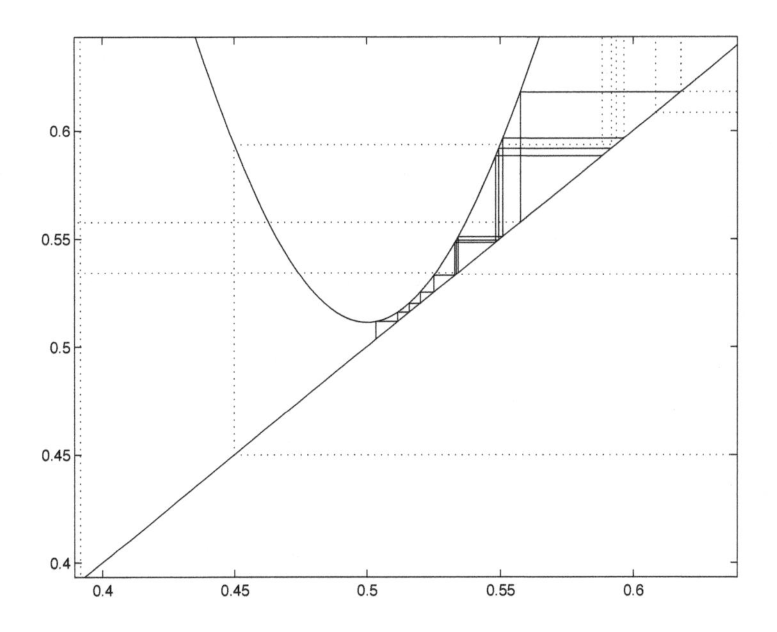

Figure 3.6: Zooming in on the central portion of the preceding figure.

Chapter 4

Conjugacy.

We now embark on a question which will occupy us a lot in this book: When are two dynamical systems "the same"?

Here is an example of what I mean:

4.1 Affine equivalence.

An **affine transformation** of the real line is a transformation of the form

$$x \mapsto h(x) = Ax + B$$

where A and B are real constants with $A \neq 0$. So an affine transformation consists of a change of scale (and possibly direction if $A < 0$) given by the factor A, followed by a shift of the origin given by B. In the study of linear phenomena, we expect that many of the essentials of an object be invariant under a change of scale and a shift of the origin of our coordinate system.

For example, consider the logistic transformation, $L_\mu(x) = \mu x(1-x)$ and the affine transformation

$$h_\mu(x) = -\mu x + \frac{\mu}{2}.$$

We claim that

$$h_\mu \circ L_\mu \circ h_\mu^{-1} = Q_c \tag{4.1}$$

where

$$Q_c(x) = x^2 + c \tag{4.2}$$

and where c is related to μ by the equation

$$c = -\frac{\mu^2}{4} + \frac{\mu}{2}. \tag{4.3}$$

In other words, we are claiming that if c and μ are related by (4.3) then we have

$$h_\mu(L_\mu(x)) = Q_c(h_\mu(x)).$$

To check this, the left hand side expands out to be

$$-\mu[\mu x(1-x)]+\frac{\mu}{2}=\mu^2x^2-\mu^2x+\frac{\mu}{2},$$

while the right hand side expands out as

$$(-\mu x+\frac{\mu}{2})^2-\frac{\mu^2}{4}+\frac{\mu}{2}=\mu^2x^2-\mu^2x+\frac{\mu}{2}$$

giving the same result as before, proving (4.1).

We say that the transformations L_μ and $Q_c, c=-\frac{\mu^2}{4}+\frac{\mu}{2}$ are **conjugate** by the affine transformation, h_μ.

4.1.1 Conjugacy in general.

More generally, let $f: X \to X$ and $g: Y \to Y$ be maps of the sets X and Y to themselves, and let $h: X \to Y$ be a one to one map of X onto Y. We say that h conjugates f into g if

$$h \circ f \circ h^{-1} = g,$$

or, what amounts to the same thing, if

$$h \circ f = g \circ h.$$

We shall frequently write this equation in the form of a *commutative diagram*

$$\begin{array}{ccc} X & \xrightarrow{f} & X \\ h\downarrow & & \downarrow h \\ Y & \xrightarrow[g]{} & Y \end{array}$$

The statement that the diagram is commutative means that going along the upper right hand path (so applying $h \circ f$) is equal to traversing the left lower path (which is $g \circ h$).

Notice that if $h \circ f \circ h^{-1} = g$, then

$$g^{\circ n} = h \circ f^{\circ n} \circ h^{-1}.$$

So the problem of studying the iterates of g is the same (up to the transformation h) as that of f, *provided* that the properties we are interested in studying are not destroyed by h.

Certainly affine transformations will always be allowed. Let us generalize (4.1) by showing that *any* quadratic transformation (with non-vanishing leading term) is conjugate (by an affine transformation) to a transformation of the form Q_c for suitable c. More precisely:

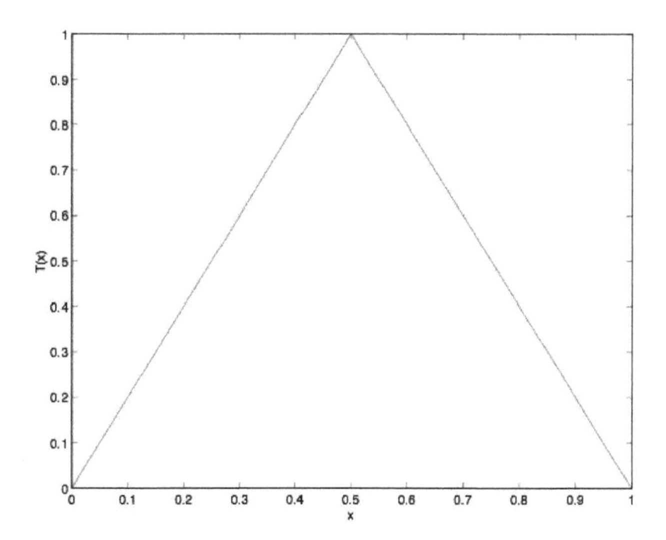

Figure 4.1: The tent transformation.

Proposition 4.1.1. *Let $f = ax^2 + bx + d$ then f is conjugate to Q_c by the affine map $h(x) = Ax + B$ where*

$$A = a, \quad B = \frac{b}{2}, \quad \textit{and} \quad c = ad + \frac{b}{2} - \frac{b^2}{4}.$$

Proof. Direct verification shows that $h \circ f = Q_c \circ h$. □

Let us understand the importance of this result. The general quadratic transformation f depends on three parameters a, b and d. But if we are interested in the qualitative behavior of the iterates of f, it suffices to examine the one parameter family Q_c. Any quadratic transformation (with non-vanishing leading term) has the same behavior (in terms of its iterates) as one of the Q_c. The family of possible behaviors under iteration is one dimensional, depending on a single parameter c. We may say that the family Q_c (or for that matter the family L_μ) is *universal* with respect to quadratic maps as far as iteration is concerned.

4.2 The tent transformation and L_4.

Let $T : [0, 1] \to [0, 1]$ be the map defined by

$$T(x) = 2x, \ 0 \le x \le \frac{1}{2}, \quad T(x) = -2x + 2, \ \frac{1}{2} \le x \le 1.$$

So the graph of T looks like a tent, hence its name. It consists of the straight line segment of slope 2 joining $x = 0, y = 0$ to $x = \frac{1}{2}, y = 1$ followed by the segment of slope -2 joining $x = \frac{1}{2}, y = 1$ to $x = 1, y = 0$.

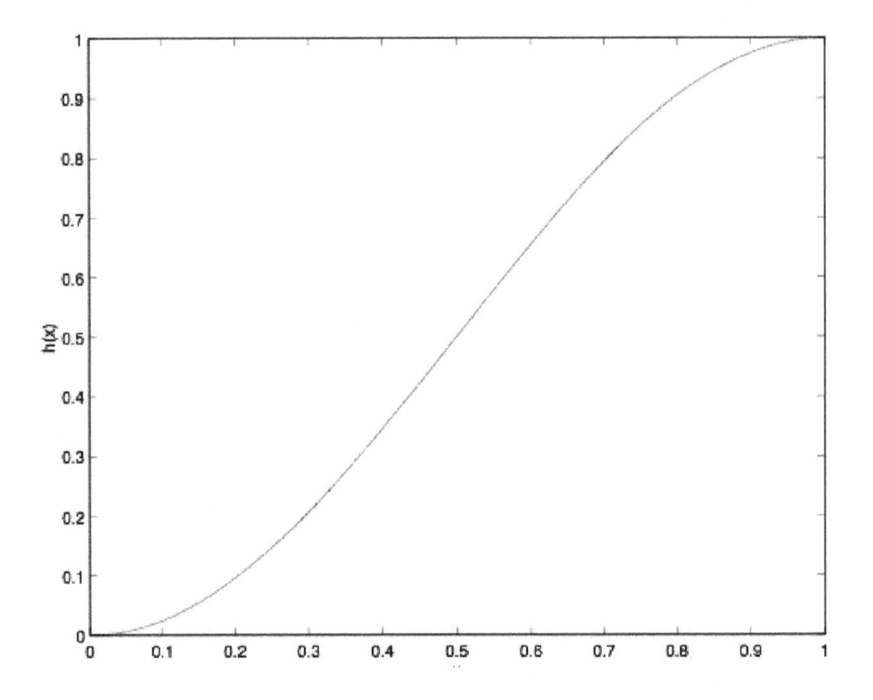

Figure 4.2: $h(x) = \sin^2\left(\frac{\pi x}{2}\right)$.

Of course, here L_4 is our old friend, $L_4(x) = 4x(1-x)$. We wish to show that

$$L_4 \circ h = h \circ T$$

where

$$h(x) = \sin^2\left(\frac{\pi x}{2}\right).$$

In other words, we claim that the our diagram above commutes when $f = T$, $g = L_4$ and h is as given. The function $\sin\theta$ increases monotonically from 0 to 1 as θ increases from 0 to $\pi/2$. So, setting

$$\theta = \frac{\pi x}{2},$$

we see that $h(x)$ increases monotonically from 0 to 1 as x increases from 0 to 1. It therefore is a one to one continuous map of $[0,1]$ onto itself, and thus has a continuous inverse. It is differentiable everywhere with $h(x) > 0$ for $0 < x < 1$. But $h'(0) = h'(1) = 0$. So h^{-1} is em not differentiable at the end points, but *is* differentiable for $0 < x < 1$. To verify our claim, we substitute

$$\begin{aligned}L_4(h(x)) &= 4\sin^2\theta(1-\sin^2\theta)\\ &= 4\sin^2\theta\cos^2\theta\\ &= \sin^2 2\theta\\ &= \sin^2\pi x.\end{aligned}$$

So for $0 \leq x \leq \frac{1}{2}$ we have verified that

$$L_4(h(x)) = h(2x) = h(T(x))$$

For $\frac{1}{2} < x \leq 1$ we have

$$\begin{aligned}h(T(x)) &= h(2-2x)\\ &= \sin^2(\pi - \pi x)\\ &= \sin^2\pi x\\ &= \sin^2 2\theta\\ &= 4\sin^2\theta(1-\sin^2\theta)\\ &= L_4(h(x))\end{aligned}$$

where we have used the fact that $\sin(\pi - \alpha) = \sin\alpha$ to pass from the second line to the third. So we have verified our claim in all cases. □

Here is another example of a conjugacy, this time an affine conjugacy. Consider

$$V(x) = 2|x| - 2.$$

V is a map of the interval $[-2, 2]$ into itself. Consider

$$h_2(x) = 2 - 4x.$$

So $h_2(0) = 2, h_2(1) = -2$. In other words, h_2 maps the interval $[0, 1]$ in a one to one fashion onto the interval $[-2, 2]$.

We claim that

$$V \circ h_2 = h_2 \circ T.$$

Indeed,

$$V(h_2(x)) = 2|2 - 4x| - 2.$$

For $0 \leq x \leq \frac{1}{2}$ this equals $2(2-4x) - 2 = 2 - 8x = 2 - 4(2x) = h_2(Tx)$. For $\frac{1}{2} \leq x \leq 1$ we have $V(h_2(x)) = 8x - 6 = 2 - 4(2 - 2x) = h_2(Tx)$. So we have verified the required equation in all cases. The effect of the affine transformation, h_2 is to enlarge the graph of T, shift it, and turn it upside down. But as far as iterations are concerned, these changes do not effect the essential behavior.

4.3 Chaos.

4.3.1 Transitivity.

A transformation F is called (topologically) **transitive** if for any two open (non empty) intervals, I and J, one can find initial values in I which, when iterated,

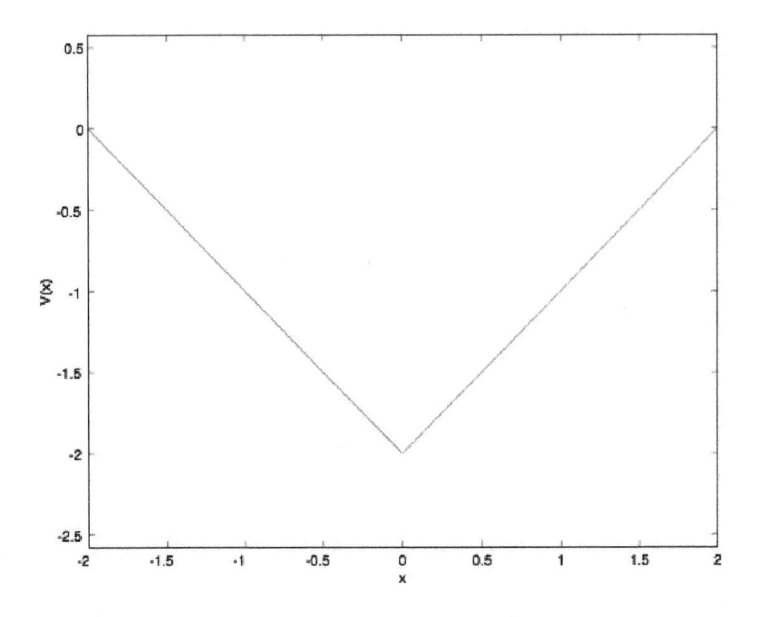

Figure 4.3: $V(x) = 2|x| - 2$.

will eventually take values in J. In other words, we can find an $x \in I$ and an integer n so that $F^n(x) \in J$.

The tent map is transitive.

For example, consider the tent transformation, T. Notice that T maps the interval $[0, \frac{1}{2}]$ onto the entire interval $[0, 1]$, and also maps the interval $[\frac{1}{2}, 1]$ onto the entire interval, $[0, 1]$. So $T^{\circ 2}$ maps each of the intervals $[0, \frac{1}{4}], [\frac{1}{4}, \frac{1}{2}], [\frac{1}{2}, \frac{3}{4}]$ and $[\frac{3}{4}, 1]$ onto the entire interval $[0, 1]$. More generally, $T^{\circ n}$ maps each of the 2^n intervals $[\frac{k}{2^n}, \frac{k+1}{2^n}]$, $0 \leq k \leq 2^n - 1$ onto the entire interval $[0, 1]$. But any open interval I contains some interval of the form $[\frac{k}{2^n}, \frac{k+1}{2^n}]$ if we choose n sufficiently large. For example it is enough to choose n so large that $\frac{3}{2^n}$ is less than the length of I. So for this value on n, $T^{\circ n}$ maps I onto the entire interval $[0, 1]$, and so, in particular, there will be points, x, in I with $F(x) \in J$.

Transitivity and conjugacy.

Proposition 4.3.1. *Suppose that $g \circ h = h \circ f$ where h is continuous and surjective, and suppose that f is transitive. Then g is transitive.*

Proof. We are given non-empty open I and J and wish to find an n and an $x \in I$ so that $g^{\circ n}(x) \in J$. To say h is continuous means that $h^{-1}(J)$ is a union of open intervals. To say that h is surjective implies that $h^{-1}(J)$ is not empty. Let L be one of the intervals constituting $h^{-1}(J)$. Similarly, $h^{-1}(I)$ is a union of open intervals. Let K be one of them. By the transitivity of f we can find an n and a $y \in K$ with $f^{\circ n}(y) \in L$. Let $x = h(y)$. Then $x \in I$ and $g^{\circ n}(x) = g^{\circ n}(h(y)) = h(f^{\circ n}(y)) \in h(L) \subset J$. □

Homeomorphisms.

Many interesting properties of a transformation are preserved under conjugation by a homeomorphism. (A **homeomorphism** is a bijective continuous map with continuous inverse.) For example, if p is a periodic point of period n of f, so that $f^{\circ n}(p) = p$, then

$$g^{\circ n}(h(p)) = h \circ f^{\circ n}(p) = h(p)$$

if $h \circ f = g \circ h$. So periodic points are carried into periodic points of the same period under a conjugacy by a homeomorphism. The previous proposition implies that if f is conjugate to g by a homeomorphism, then f is transitive if and only if g is transitive.

We will consider several other important properties of a transformation as we go along, and will prove that they are invariant under such a conjugacy. So what our result means is that if we prove these properties for T, we conclude that they are true for L_μ. Since we have verified that L_4 is conjugate to Q_{-2} and V, we conclude that they hold for Q_{-2} and V as well.

4.3.2 Density of periodic points.

A set S of points is called **dense** if every non-empty open interval, I, contains a point of S. The behavior of density under continuous surjective maps is also very simple:

Proposition 4.3.2. *If $h : X \to Y$ is a continuous surjective map, and if D is a dense subset of X then $h(D)$ is a dense subset of Y.*

Proof. Let $I \subset Y$ be a non-empty open interval. Then $h^{-1}(I)$ is a union of open intervals. Pick one of them, K and then a point $y \in D \cap K$ which exists since D is dense. But then $f(y) \in f(D) \cap I$. □

We define PER(f) to be the set of periodic points of the map f (including the fixed points). If $h \circ f = g \circ h$, then $f^{\circ n}(p) = p$ implies that $g^{\circ n}(h(p)) = h(f^{\circ n}(p)) = h(p)$ so

$$h[\mathrm{PER}(f)] \subset \mathrm{PER}(g).$$

In particular, if h is continuous and surjective, and if PER(f) is dense, then so is PER(g).

4.3.3 A definition of chaos.

There are various mathematical definitions of the popular word "chaotic". We will pick one:

Following Devaney and J. Banks et.al. *Amer. Math. Monthly* **99** (1992) 332-334, let us call f **chaotic** if f is transitive and PER(f) is dense. It follows from the above discussion that

Proposition 4.3.3. *Let $h : X \to Y$ is surjective and continuous. If $f : X \to X$ is chaotic, and if $h \circ f = g \circ h$, then g is chaotic.*

Important remark. Notice that the proposition does *not* require that h^{-1} exists or, if it does, that h^{-1} be continuous.

The tent map is chaotic.

We have already verified that the tent transformation, T, is transitive. We claim that PER(T) is dense on $[0,1]$ and hence that T is chaotic. To see this, observe that T^n maps the interval $[\frac{k}{2^n}, \frac{k+1}{2^n}]$ onto $[0,1]$. In particular, there is a point $x \in [\frac{k}{2^n}, \frac{k+1}{2^n}]$ which is mapped into itself. In other words, every interval $[\frac{k}{2^n}, \frac{k+1}{2^n}]$ contains a periodic point for T. But any non-empty open interval I contains an interval of the type $[\frac{k}{2^n}, \frac{k+1}{2^n}]$ for sufficiently large n. Hence T is chaotic.

From the above propositions it follows that L_4, Q_{-2}, and V are all chaotic.

4.3.4 The sawtooth transformation and the shift.

The sawtooth transformation.

Define the **sawtooth function** S by

$$S(x) = 2x, \quad 0 \leq x < \frac{1}{2}, \qquad S(x) = 2x - 1, \quad \frac{1}{2} \leq x \leq 1. \tag{4.4}$$

The sawtooth map and the tent map.

The map S is discontinuous at $x = .5$. However, we can find a continuous, surjective map, h, such that $h \circ S = T \circ h$. In fact, we can take h to be T itself! In other words we claim that

$$\begin{array}{ccc} I & \xrightarrow{S} & I \\ {\scriptstyle T}\downarrow & & \downarrow{\scriptstyle T} \\ I & \xrightarrow[T]{} & I \end{array}$$

commutes where $I = [0,1]$.

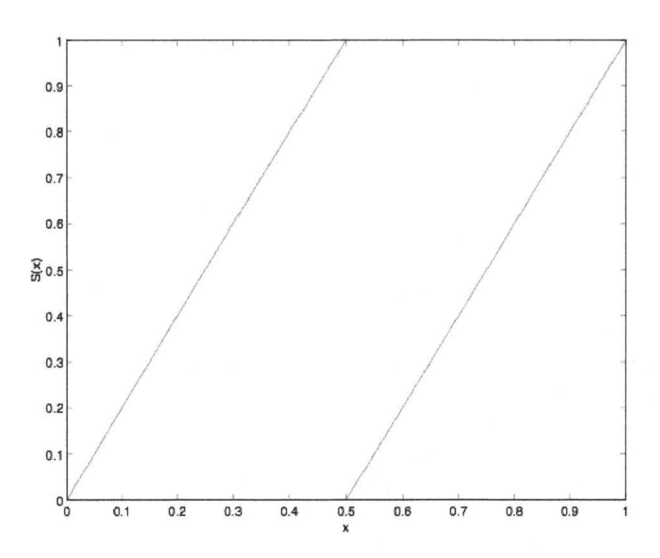

Figure 4.4: The sawtooth function S.

To verify this, we successively compute both $T \circ T$ and $T \circ S$ on each of the quarter intervals:

$$
\begin{array}{rclcll}
T(T(x)) & = & T(2x) & = & 4x & \text{for } 0 \leq x \leq 0.25 \\
T(S(x)) & = & T(2x) & = & 4x & \text{for } 0 \leq x \leq 0.25 \\
T(T(x)) & = & T(2x) & = & -4x+2 & \text{for } 0.25 < x < 0.5 \\
T(S(x)) & = & T(2x) & = & -4x+2 & \text{for } 0.25 \leq x < 0.5 \\
T(T(x)) & = & T(-2x+2) & = & 4x-2 & \text{for } 0.5 \leq x \leq 0.75 \\
T(S(x)) & = & T(2x-1) & = & 4x-2 & \text{for } 0.5 \leq x \leq 0.75 \\
T(T(x)) & = & T(-2x+2) & = & -4x+4 & \text{for } 0.75 < x \leq 1 \\
T(S(x)) & = & T(2x-1) & = & -4x+4 & \text{for } 0.75 < x \leq 1
\end{array}
$$

The h that we are using (namely $h = T$) is not one to one. That is why our diagram can commute even though T is continuous and S is not.

The one-sided shift.

Let X be the set of infinite (one sided) sequences of zeros and ones. So a point of X is a sequence $\{a_1 a_2 a_3 \dots\}$ where each a_i is either 0 or 1. However we exclude all points with a tail consisting of infinite repeating $1'$s. So a sequence such as $\{00111111111\dots\}$ is excluded. We will identify X, as a *set*, with the half open interval $[0, 1)$ by assigning to each point $x \in [0, 1)$ its binary expansion, where we agree that all points of the form $\frac{k}{2^n}$ have binary expansions ending in zeros. (See below for further details).

Conversely, we assign to each sequence $a = \{a_1a_2a_3\dots\}$ the number

$$h(a) = \sum \frac{a_i}{2^i}.$$

The map

$$h : X \to [0,1)$$

just defined is clear. The inverse map, assigning to each real number between 0 and 1 its binary expansion deserves a little more discussion: Take a point $x \in [0,1)$. If $x < \frac{1}{2}$ the first entry in its binary expansion is 0. If $\frac{1}{2} \leq x$ then the first entry in the binary expansion of x is 1. Now apply S. If $S(x) < \frac{1}{2}$ (which means that either $0 \leq x < \frac{1}{4}$ or $\frac{1}{2} \leq x < \frac{3}{4}$) then the second entry of the binary expansion of x is 0, while if $\frac{1}{2} \leq S(x) < 1$ then the second entry in the binary expansion of x is 1. Thus the operator S provides the algorithm for the computation of the binary expansion of x.

Let us consider, for example, $x = \frac{7}{16}$. Then the sequence $\{S^k(x)\}, k = 0,1,2,3,\dots$ is

$$\frac{7}{16}, \frac{7}{8}, \frac{3}{4}, \frac{1}{2}, 0, 0, 0, \dots.$$

In general it is clear that for any number of the form $\frac{k}{2^n}$, after $n-1$ iterations of the operator S the result will be either 0 or $\frac{1}{2}$. So all $S^k(x) = 0, k \geq n$. In particular, no infinite sequence with a tail of repeating 1's can arise. We see that the binary expansion of $h(a)$ gives us a back, so we may (and shall) identify X with $[0,1)$.

A topology on X.

Notice that we did not start with any independent notion of topology or metric on X. But the map $h : X \to [0,1)$, suggests a notion of distance on X: For example, if the binary expansions of x and y agree up to the kth position, then

$$|x - y| < 2^{-k}.$$

So we *define* the distance $d(a,b)$ between two sequences a and b to be 2^{-k} where k is the first place they do not agree. (Of course we define the distance from an a to itself to be zero.)

The map h is continuous.

If $h(a) = x$ and $|y - x| < \epsilon$, choose k such that $2^{-k} < \epsilon$. If $d(a,b) < 2^{-k}$ then $h(b)$ has the same binary expansion as x up to order k which implies that $|x - h(b)| < \epsilon$.

The map h^{-1}, although it exists, is not continuous: If we take $x = .1000000\cdots$ and $y = .011111111\cdots1000000\cdots$ then $|x-y|$ can be made as small as we like by choosing a large enough collection of 1's. But $d(h^{-1}(x), h^{-1}(y)) = \frac{1}{2}$.

The shift.

Consider the map $\mathbf{Sh}: X \to X$ defined as follows:

$$\mathbf{Sh}(a_1a_2a_3a_4\ldots) := a_2a_3a_4a_5\ldots.$$

In other words, **Sh** consists of lopping off the first entry of a and shifting all the rest one unit to the left. For this reason it is called the **shift map**.

If $d(a,b) < 2^{-k-1}$ then $d(\mathbf{Sh}(a), \mathbf{Sh}(b)) < 2^{-k}$, showing that **Sh** is continuous.

The periodic points of **Sh** are very easy to describe: A point a is periodic with period n under the shift map if and only if it consists of a repeating finite sequence of length n. A point of period three, for example, has the form

$$a_1a_2a_3a_1a_2a_3a_1a_2a_3\cdots.$$

The shift map is chaotic.

Given $a \in X$ we can find a periodic point (of period k) within distance 2^{-k} by simply taking the first k entries in a and then repeating them indefinitely. This shows that PER(**Sh**) is dense in X.

Let J be a set containing all points c of distance less than 2^{-n} about a point b, and let I be a set containing all points of distance less than 2^{-n} about a point a. Consider the point

$$c = a_1a_2\cdots a_na_{n+1}b_1b_2b_3b_4\cdots.$$

Then $c \in I$ and $\mathbf{Sh}^{\circ(n+1)} = b$. This shows that **Sh** is transitive.

Conclusion: Sh is chaotic.

Confession: In the above discussion I have generalized the concepts involving chaos from an interval to a metric space. I hope that a formal redefinition is not necessary.

Back to the sawtooth map.

The expression of the sawtooth map S in terms of the binary representation is very simple:

$$S : .a_1a_2a_3a_4\ldots \mapsto .a_2a_3a_4a_5\ldots.$$

It consists of throwing away the first digit and then shifting the entire sequence one unit to the left.

The map $h: X \to [0,1)$ consists of putting a period in front of the sequence a. This shows that

$$S \circ h = h \circ \mathbf{Sh}.$$

In other words, we have the commutative diagram

$$\begin{array}{ccc} X & \xrightarrow{\mathbf{Sh}} & X \\ h\downarrow & & \downarrow h \\ [0,1) & \xrightarrow[S]{} & [0,1) \end{array}$$

showing that S is chaotic on $[0,1)$.

Of course, once we know that S is chaotic on the open interval $[0,1)$, we know that it is chaotic on the closed interval $[0,1]$ since the addition of one extra point (which gets mapped to 0 by S) does not change the requirements of being chaotic.

Going to the unit circle.

Now consider the map $t \mapsto e^{2\pi it}$ of $[0,1]$ onto the unit circle, S^1. Another way of writing this map is to describe a point on the unit circle by $e^{i\theta}$ where θ is an angular variable, that is θ and $\theta + 2\pi$ are identified. Then the map is $t \mapsto 2\pi t$. This map, h, is surjective and continuous and is one to one except at the end points: 0 and 1 are mapped into the same point of S^1.

Clearly

$$h \circ S = D \circ h$$

where

$$D(\theta) = 2\theta.$$

Or, if we write $z = e^{i\theta}$, then in terms of z, the map D sends

$$z \mapsto z^2.$$

So D is called the **doubling** map or the squaring map. We have proved that it is chaotic.

The doubling map and Q_{-2}.

We can use the fact that D is chaotic to give an alternative proof of the fact that Q_{-2} is chaotic. Indeed, consider the map $h : S^1 \to [-2,2]$

$$h(\theta) = 2\cos\theta.$$

It is clearly surjective and continuous. We claim that

$$h \circ D = Q_{-2} \circ h.$$

Indeed,

$$h(D(\theta)) = 2\cos 2\theta = 2(2\cos^2\theta - 1) = (2\cos\theta)^2 - 2 = Q_{-2}(h(\theta)).$$

This gives an alternative proof that Q_{-2} (and hence L_4 and T) are chaotic.

4.4 Sensitivity to initial conditions

In this section we prove that if f is chaotic, then f is sensitive to initial conditions in the sense of the following:

Proposition 4.4.1. **([Sensitivity.])** *Let $f : X \to X$ be a chaotic transformation. Then there is an $\delta > 0$ such that for any $x \in X$ and any open set J containing x and some points other than x, there is a point $y \in J$ and an integer, n with*

$$d(f^{\circ n}(x), f^{\circ n}(y)) > \delta. \tag{4.5}$$

In other words, we can find points arbitrarily close to x which move a distance at least d away under some interation of f. This for any $x \in X$. For the proof, we begin with a lemma.

Lemma 4.4.1. *Let $f : X \to X$ be a transformation with at least two distinct periodic orbits. There is a $c > 0$ with the property that for any $x \in X$ there is a periodic point p such that*

$$d(x, f^{\circ k}(p)) > c, \quad \forall k.$$

Proof of the lemma. Choose two periodic points, r and s with distinct orbits, so that $d(f^{\circ k}(r), f^{\circ \ell}(s)) > 0$ for all k and ℓ. Choose c so that $2c < \min d(f^{\circ k}(r), f^{\circ \ell}(s))$. Then for all k and ℓ we have

$$\begin{aligned} 2c &< d(f^{\circ k}(r), f^{\circ \ell}(s)) \\ &\leq d(f^{\circ k}(r), x) + d(x, f^{\circ \ell}(s)) \quad \text{by the triangle inequality.} \end{aligned}$$

If x is within distance c of *any* of the points $f^{\circ \ell}(s)$ then it must be at a greater distance than c from *all* of the points $f^{\circ k}(r)$ and similarly, if x is within distance c of *any* of the points $f^{\circ k}(r)$ it must be at a greater distance than c from *all* of the points $f^{\circ \ell}(s)$. So one of the two, (or both) of the r or s will work as the p for x. □

Proof of Proposition 4.4.1 with $\delta = c/4$. Let x be any point of X and J any open set containing x. Since the periodic points of f are dense, we can find a periodic point q of f in

$$U = J \cap B_d(x),$$

where $B_\delta(x)$ denotes the open "ball" of length r centered at x,

$$B_\delta(x) = \{y \in X | d(x, y) < \delta\}.$$

Let n be the period of q. Let p be a periodic point whose orbit is of distance greater than 4δ from x, and set

$$W_i = B_\delta(f^{\circ i}(p)).$$

Since $f^{\circ i}(p) \in W_i$, i.e. $p \in f^{-i}(W_i) := (f^{\circ i})^{-1}(W_i)$ for all i, we see that the open set $V := f^{-1}(W_1) \cap f^{-2}(W_2) \cap \cdots \cap f^{-n}(W_n)$ is not empty.

Now we use the transitivity property of f applied to the open sets U and V. By assumption, we can find a $z \in U$ and a positive integer k such that $f^{\circ k}(z) \in V$. Let j be the smallest integer so that $k < nj$. In other words,

$$1 \leq nj - k \leq n.$$

So

$$f^{\circ nj}(z) = f^{\circ(nj-k)}(f^{\circ k}(z)) \in f^{\circ(nj-k)}(V).$$

But

$$\begin{aligned} f^{\circ(nj-k)}(V) &= f^{\circ(nj-k)}\left(f^{-1}(W_1) \cap f^{-2}(W_2) \cap \cdots \cap f^{-n}(W_n)\right) \\ &\subset f^{\circ(nj-k)}(f^{-(nj-k)}W_{nj-k}) \\ &= W_{nj-k}. \end{aligned}$$

In other words, $\quad d(f^{nj}(z), f^{nj-k}(p)) < \delta.$

On the other hand, $f^{nj}(q) = q$, since n is the period of q. Thus

$$d(f^{nj}(q), f^{nj}(z)) = d(q, f^{nj}(z))$$

$$\geq d(x, f^{\circ(nj-k)}(p)) - d(f^{\circ(nj-k)}(p), f^{nj}(z)) - d(q, x)$$

by the triangle inequality since

$$d(q, x) \leq d(x, f^{\circ(nj-k)}(p)) + d(f^{\circ(nj-k)}(p), f^{\circ nj}(z)) + d(q, f^{\circ nj}(z)).$$

So

$$d(f^{nj}(q), f^{nj}(z)) \geq 4\delta - \delta - \delta = 2\delta.$$

But this inequality implies that either

$$d(f^{\circ nj}(x), f^{\circ nj}(z)) \geq \delta$$

or

$$d(f^{\circ nj}(x), f^{\circ nj}(q) \geq \delta$$

for if $f^{\circ nj}(x)$ were within distance $< \delta$ from both of these points, they would have to be within distance $< 2\delta$ from each other, contradicting the top inequality above. So one of the two, z or q will serve as the y in the proposition with $m = nj$.

Philosophical implications.

Much is made in the popular literature of the property of sensitivity to initial conditions having to do with issues of *determinism.* The idea is that a transformation like L_4 is **"bad"** because it is unpredictable in the sense a small change in the "initial conditions" implies that after a while, there is a large deviation in the final results. This is known as the "butterfly effect".

While L_4 may be "bad" in this sense, it does have very nice statistical properties as will be explained in the next chapter. So as is frequently true in life, the nature of the answer depends on how the question is posed.

4.5 Conjugacy for monotone maps.

We begin this section by showing that if f and g are continuous strictly monotone maps of the unit interval $I = [0,1]$ onto itself, and if their graphs are both strictly below (or both strictly above) the line $y = x$ in the interior of I, then they are conjugate by a homeomorphism. Here is the precise statement:

Proposition 4.5.1. *Let f and g be two continuous strictly increasing functions defined on $[0,1]$ and satisfying*

$$\begin{aligned} f(0) &= 0 \\ g(0) &= 0 \\ f(1) &= 1 \\ g(1) &= 1 \\ f(x) &< x \quad \forall x \neq 0,1 \\ g(x) &< x \quad \forall x \neq 0,1. \end{aligned}$$

Then there exists a continuous, monotone increasing function h defined on $[0,1]$ with

$$h(0) = 0, \qquad h(1) = 1,$$

and

$$h \circ f = g \circ h.$$

Proof. Choose any point (x_0, y_0) in the open square

$$0 < x < 1, \;\; 0 < y < 1.$$

If (x_0, y_0) is to be a point on the curve $y = h(x)$, then the equation $h \circ f = g \circ h$ implies that the point (x_1, y_1) also lies on this curve, where

$$x_1 = f(x_0), \quad y_1 = g(y_0).$$

By induction so will the points (x_n, y_n) where

$$x_n = f^{\circ n}(x_0), \;\; y_n = g^{\circ n}(y_0).$$

By hypothesis

$$x_0 > x_1 > x_2 > ...,$$

and since there is no solution to $f(x) = x$ for $0 < x < 1$ the limit of the x_n, as $n \to \infty$ must be zero. Also for the y_n. So the sequence of points (x_n, y_n) approaches $(0, 0)$ as $n \to +\infty$. Similarly, as $n \to -\infty$ the points (x_n, y_n) approach $(1, 1)$. Now choose any continuous, strictly monotone function

$$y = h(x),$$

defined on

$$x_1 \leq x \leq x_0$$

with

$$h(x_1) = y_1, \quad h(x_0) = y_0.$$

Extend its definition to the interval $x_2 \leq x \leq x_1$ by setting

$$h(x) = g(h(f^{-1}(x))), \quad x_2 \leq x \leq x_1.$$

Notice that at x_1 we have

$$g(h(f^{-1}(x_1))) = g(h(x_0)) = g(y_0) = y_1,$$

so the definitions of h at the point x_1 are consistent. Since f and g are monotone and continuous, and since h was chosen to be monotone on $x_1 \leq x \leq x_0$, we conclude that h is monotone on $x_2 \leq x \leq x_1$ and hence continuous and monotone on all of $x_2 \leq x \leq x_0$. Continuing in this way, we define h on the interval $x_{n+1} \leq x \leq x_n$, $n \geq 0$ by

$$h = g^n \circ h \circ f^{-n}.$$

Setting $h(0) = 0$, we get a continuous and monotone increasing function defined on $0 \leq x \leq x_0$. Similarly, we extend the definition of h to the right of x_0 up to $x = 1$. By its very construction, the map h conjugates f into g, proving the proposition.

□

Notice that as a corollary of the method of proof, we can conclude

Corollary 4.5.1. *Let f and g be two monotone increasing functions defined in some neighborhood of the origin and satisfying*

$$f(0) = g(0) = 0, \quad |f(x)| < |x|, \ |g(x)| < |x|, \ \forall x \neq 0.$$

Then there exists a homeomorphism, h defined in some neighborhood of the origin with $h(0) = 0$ and

$$h \circ f = g \circ h.$$

Indeed, just apply the method (for $n \geq 0$) to construct h to the right of the origin, and do an analogous procedure to construct h to the left of the origin. As a special case we obtain

Theorem 4.5.1. *Let f and g be differentiable functions with $f(0) = g(0) = 0$ and*

$$0 < f'(0) < 1, \quad 0 < g'(0) < 1. \tag{4.6}$$

Then there exists a homeomorphism h defined in some neighborhood of the origin with $h(0) = 0$ and which conjugates f into g.

Proof. The mean value theorem guarantees that the hypotheses of the preceding corollary are satisfied. □

I is clear that we can replace (4.6) by any of the conditions

$$\begin{array}{ll} 1 < f'(0), & 1 < g'(0) \\ 0 > f'(0) > -1, & 0 > g'(0) > -1 \\ -1 > f'(0), & -1 > g'(0), \end{array}$$

and the conclusion of the theorem still holds.

It is important to observe that if $f'(0) \neq g'(0)$, then the homeomorphism, h, can not be a diffeomorphism. That is, h can not be differentiable with a differentiable inverse. In fact, h can not have a non-zero derivative at the origin. Indeed, differentiating the equation $g \circ h = h \circ f$ at the origin gives

$$g'(0)h'(0) = h'(0)f'(0),$$

and if $h'(0) \neq 0$ we can cancel it form both sides of the equation so as to obtain

$$f'(0) = g'(0). \tag{4.7}$$

What *is* true is that if (4.7) holds, and if

$$|f'(0)| \neq 1, \tag{4.8}$$

then we can find a differentiable h with a differentiable inverse which conjugates f into g.

These theorems are among my earliest mathematical theorems. A complete characterization of transformations of $\mathbb{R}$ near a fixed point together with the conjugacy by smooth maps if (4.7) and (4.8) hold, were obtained and submitted for publication in 1955 and published in the Duke Mathematical Journal. The discussion of equivalence under homeomorphism or diffeomorphism in n-dimensions was treated for the case of contractions in 1957 and in the general case in 1958, both papers appearing in the American Journal of Mathematics. We will return to these matters later.

4.6 Sequence space and symbolic dynamics.

In this section we will illustrate a powerful method for studying dynamical systems by examining the quadratic transformation

$$Q_c : x \mapsto x^2 + c$$

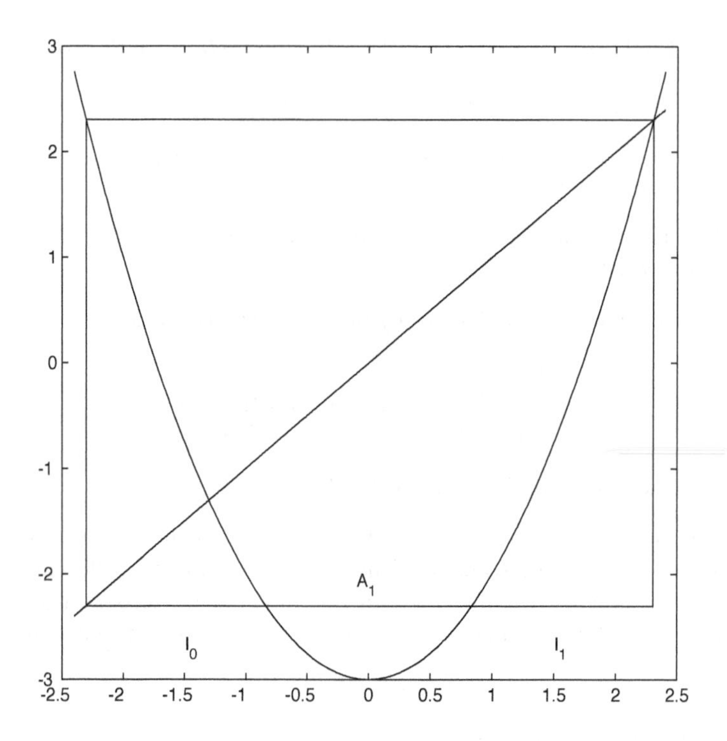

Figure 4.5: Q_3.

for values of $c < -2$.

For *any* value of c, the two possible fixed points of Q_c are

$$p_-(c) = \frac{1}{2}(1 - \sqrt{1-4c}), \quad p_+(c) = \frac{1}{2}(1 + \sqrt{1-4c})$$

by the quadratic formula. These roots are real with $p_-(c) < p_+(c)$ for $c < 1/4$.

The graph of Q_c lies above the diagonal for $x > p_+(c)$, hence the iterates of any $x > p_+(c)$ tend to $+\infty$. If $x_0 < -p_+(c)$, then $x_1 = Q_c(x_0) > p_+(c)$, and so the further iterates also tend to $+\infty$. Hence all the interesting action takes place in the interval $[-p_+, p_+]$. The function Q_c takes its minimum value, c, at $x = 0$, and

$$c = -p_+(c) = -\frac{1}{2}(1 + \sqrt{1-4c})$$

when $c = -2$. For $-2 \leq c \leq 1/4$, the iterate of any point in $[-p_+, p_+]$ remains in the interval $[-p_+, p_+]$. But for $c < -2$ some points will escape, and it is this latter case that we want to study. o visualize the what is going on, draw the square whose vertices are at $(\pm p_+, \pm p_+)$ and the graph of Q_c over the interval $[-p_+, p_+]$. The bottom of the graph will protrude below the bottom of the square. Let A_1 denote the open interval on the x-axis (centered about the

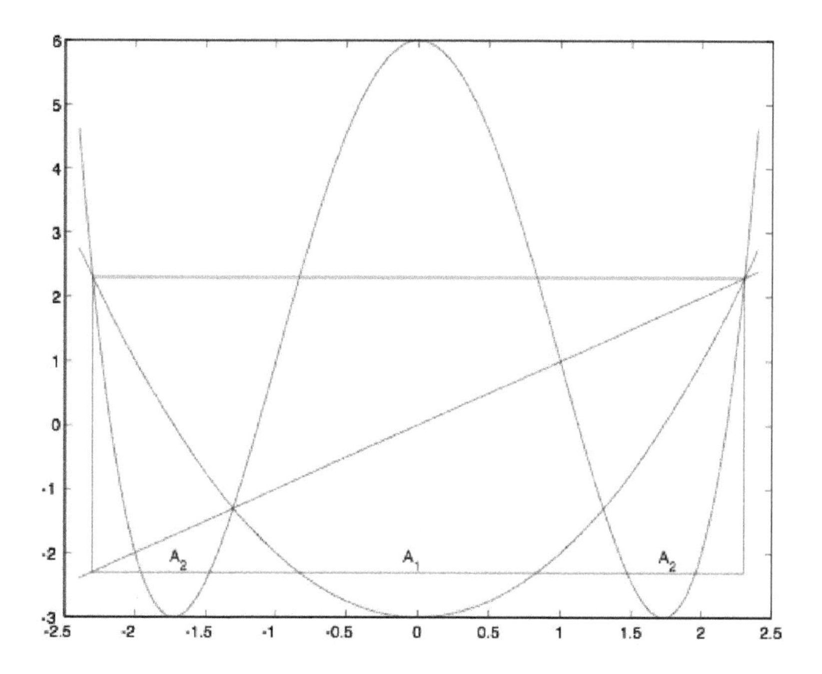

Figure 4.6: Q_3 and $Q_3^{\circ 2}$.

origin) which corresponds to this protrusion. So

$$A_1 = \{x | Q_c(x) < -p_+(c)\}.$$

Every point of A_1 escapes from the interval $[-p_+, p_+]$ after one iteration.

Let

$$A_2 = Q_c^{-1}(A_1).$$

Since every point of $[-p_+, p_+]$ has exactly two pre-images under Q_c, we see that A_2 is the union of two open intervals. To fix notation, let

$$I = [-p_+, p_+]$$

and write

$$I \backslash A_1 = I_0 \cup I_1$$

where I_0 is the closed interval to the left of A_1 and I_1 is the closed interval to the right of A_1.Thus A_2 is the union of two open intervals, one contained in I_0 and the other contained in I_1. Notice that a point of A_2 escapes from $[-p_+, p_+]$ in exactly two iterations: one application of Q_c moves it into A_1 and another application moves it out of $[-p_+, p_+]$. Conversely, suppose that a point x escapes from $[-p_+, p_+]$ in exactly two iterations. After one iteration it must lie in A_1, since these are exactly the points that escape in one iteration. Hence it must lie in A_2.

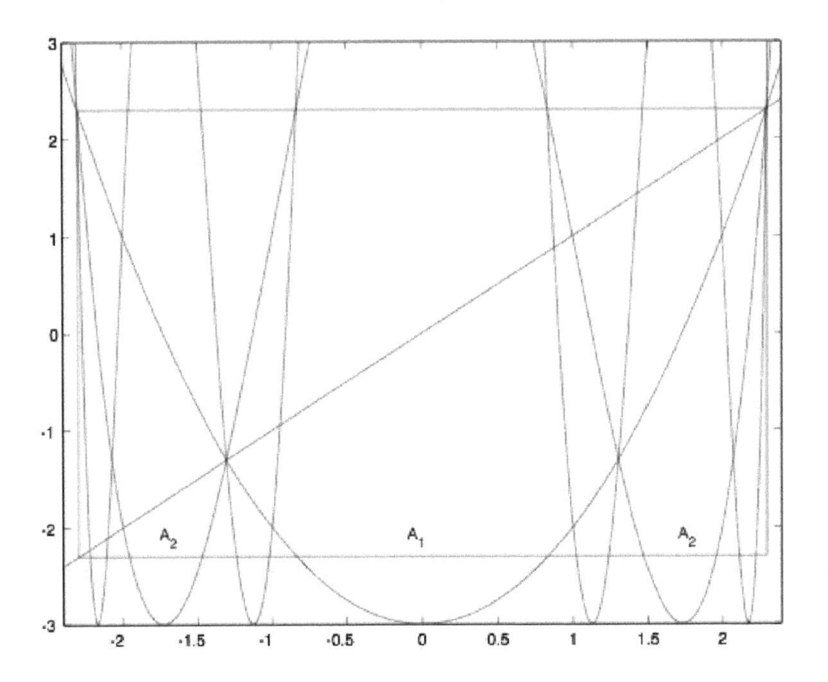

Figure 4.7: Q_3, $Q_3^{\circ 2}$ and $Q_3^{\circ 3}$.

The escapees in three or less iterations are pictured in Figure 4.7. In general, let

$$A_{n+1} = Q_c^{-\circ n}(A_1).$$

Then A_{n+1} is the union of 2^n open intervals and consists of those points which escape from $[-p_+, p_+]$ in exactly $n+1$ iterations. If the iterates of a point x eventually escape from $[-p_+, p_+]$, there must be some $n \geq 1$ so that $x \in A_n$. In other words,

$$\bigcup_{n \geq 1} A_n$$

is the set of points which eventually escape. The remaining points, those lying in the set

$$\Lambda := I \setminus \bigcup_{n \geq 1} A_n,$$

are the points whose iterates remain in $[-p_+, p_+]$ forever. The thrust of the rest of this section, is to study Λ and the action of Q_c on it.

Λ is closed and is not empty.

Since Λ is defined as the complement of an open set, we see thatΛ *is closed.* Let us show thatΛ *is not empty.* Indeed, the fixed points, $p_\pm$ certainly belong to Λ and hence so do all of their inverse images, $Q_c^{-n}(p_\pm)$.

Λ is totally disconnected.

Next we will prove

Theorem 4.6.1. *If*

$$c < -\frac{5+2\sqrt{5}}{4} \doteq -2.368\ldots \tag{4.9}$$

then Λ is totally disconnected, that is, it contains no interval.

In fact, the theorem is true for all $c < -2$ but, following [Devaney(1992)] I will only present the simpler proof when we assume (4.9). For this we use

Lemma 4.6.1. *If (4.9) holds then there is a constant $\lambda > 1$ such that*

$$|Q_c'(x)| > \lambda > 1, \quad \forall x \in I \backslash A_1. \tag{4.10}$$

Proof of the Lemma. We have $|Q_c'(x)| = |2x| > \lambda > 1$ if $|x| > \frac{1}{2}\lambda$ for all $x \in I\backslash A_1$. So we need to arrange that A_1 contains the interval $[-\frac{1}{2}, \frac{1}{2}]$ in its interior. In other words, we need to be sure that

$$Q_c(\frac{1}{2}) < -p_+.$$

The equality

$$Q_c(\frac{1}{2}) = -p_+$$

translates to

$$\frac{1}{4} + c = -\frac{1+\sqrt{1-4c}}{2}.$$

Solving the quadratic equation gives

$$c = -\frac{5+2\sqrt{5}}{4}$$

as the lower root. Hence if (4.9) holds, $Q_c(\frac{1}{2}) < -p_+$. □

Proof of Theorem 4.6.1. Suppose that there is an interval, J, contained in Λ. Then J is contained either in I_0 or I_1. In either event the map Q_c is one to one on J and maps it onto an interval. For any pair of points, x and y in J, the mean value theorem implies that

$$|Q_c(x) - Q_c(y)| > \lambda |x - y|.$$

Hence if d denotes the length of J, then $Q_c(J)$ is an interval of length at least λd contained inΛ . By induction we conclude that Λ contains an interval of length $\lambda^n d$ which is ridiculous, since eventually $\lambda^n d > 2p_+$ which is the length of I. □

Proof of the theorem. Suppose that there is an interval, J, contained inΛ. Then J is contained either in I_0 or I_1. In either event the map Q_c is one to one on J and maps it onto an interval. For any pair of points, x and y in J, the mean value theorem implies that

$$|Q_c(x) - Q_c(y)| > \lambda|x - y|.$$

Hence if d denotes the length of J, then $Q_c(J)$ is an interval of length at least λd contained inΛ . By induction we conclude that Λ contains an interval of length $\lambda^n d$ which is ridiculous, since eventually $\lambda^n d > 2p_+$ which is the length of I. □ Now consider a point $x \in \Lambda$. Either it lies in I_0 or it lies in I_1. Let us define

$$s_0(x) = 0 \;\; \forall x \in I_0$$

and

$$s_0(x) = 1 \;\; \forall x \in I_1.$$

Since all points $Q_c^{\circ n}(x)$ are inΛ , we can define $s_n(x)$ to be 0 or 1 according to whether $Q_c^{\circ n}(x)$ belongs to I_0 or I_1. In other words, we define

$$s_n(x) := \begin{cases} 0 & \text{if} \quad Q_c^{\circ n}(x) \in I_o \\ 1 & \text{if} \quad Q_c^{\circ n}(x) \in I_1 \end{cases} . \tag{4.11}$$

4.6.1 A new sequence space.

So let us introduce the **sequence space,**Σ , defined as

$$\Sigma = \{(s_0 s_1 s_2 \dots) \mid s_j = 0 \text{ or } 1\}.$$

Notice that in contrast to the space X we introduced earlier, we are not excluding any sequences.

A metric on Σ.

Define the notion of *distance* or *metric* on Σ by defining the distance between two points

$$\mathbf{s} = (s_0 s_1 s_2 \dots)$$

and

$$\mathbf{t} = (t_0 t_1 t_2 \dots)$$

to be

$$d(\mathbf{s}, \mathbf{t}) \stackrel{def}{=} \sum_{i=0}^{\infty} \frac{|s_i - t_i|}{2^i}.$$

It is immediate to check that d satisfies all the requirements for a metric: It is clear that $d(\mathbf{s}, \mathbf{t}) \geq 0$ and $d(\mathbf{s}, \mathbf{t}) = 0$ implies that $|s_i - t_i| = 0$ for all i, and hence that $\mathbf{s} = \mathbf{t}$. The definition is clearly symmetric in $\mathbf{s}$ and $\mathbf{t}$. And the usual triangle inequality

$$|s_i - u_i| \leq | s_i - t_i| + |t_i - u_i|$$

for each i implies the triangle inequality

$$d(\mathbf{s},\mathbf{u}) \leq d(\mathbf{s},\mathbf{t}) + d(\mathbf{t},\mathbf{u}).$$

Notice that if $s_i = t_i$ for $i = 0, 1, \ldots, n$ then

$$d(\mathbf{s},\mathbf{t}) = \sum_{j=n+1}^{\infty} \frac{|s_j - t_j|}{2^j} \leq \sum_{j=n+1}^{\infty} \frac{1}{2^j} = \frac{1}{2^n}.$$

Conversely, if $s_i \neq t_i$ for some $i \leq n$ then

$$d(\mathbf{s},\mathbf{t}) \geq \frac{1}{2^j} \geq \frac{1}{2^n}.$$

So if

$$d(\mathrm{s},\mathrm{t}) < \frac{1}{2^n}$$

then $s_i = t_i$ for all $i \leq n$.

4.6.2 The itinerary map.

Getting back toΛ , define the map

$$\iota : \Lambda \to \Sigma$$

by

$$\iota(x) = (s_0(x)s_1(x)s_2(x)s_3(x)\ldots) \tag{4.12}$$

where the $s_i(x)$ are defined by (4.11).

The point $\iota(x)$ is called the **itinerary** of the point x.

For example, the fixed point, p_+ lies in I_1 and hence do all of its images under Q_c^n since they all coincide with p_+. Hence its itinerary is

$$\iota(p_+) = (111111\ldots).$$

The point $-p_+$ is carried into p_+ under one application of Q_c and then stays there forever. Hence its itinerary is

$$\iota(-p_+) = (01111111\ldots).$$

The itenerary map and the shift.

It follows from the very definition that

$$\iota(Q_c(x)) = S(\iota(x))$$

where S is our old friend, the shift map,

$$S : (s_0s_1s_2s_3\ldots) \mapsto (s_1s_2s_3s_4\ldots)$$

applied to the spaceΣ . In other words,

$$\iota \circ Q_c = S \circ \iota.$$

The map ι conjugates Q_c, acting on Λ into the shift map, acting onΣ . To show that this is a legitimate conjugacy, we must prove that ι is a homeomorphism. That is, we must show that ι is one-to one, that it is onto, that it is continuous, and that its inverse is continuous:

One-to one:

Suppose that $\iota(x) = \iota(y)$ for $x, y \in \Lambda$. This means that $Q_c^n(x)$ and $Q^n(y)$ always lie in the same interval, I_0 or I_1. Thus the interval $[x, y]$ lies entirely in either I_0 or I_1 and hence Q_c maps it in one to one fashion onto an interval contained in either I_0 or I_1. Applying Q_c once more, we conclude that Q_c^2 is one-to-one on $[x, y]$. Continuing, we conclude that Q_c^n is one-to-one on the interval $[x, y]$, and we also know that (4.9) implies that the length of $[x, y]$ is increased by a factor of λ^n. This is impossible unless the length of $[x, y]$ is zero, i.e. $x = y$. □

Onto:

We start with a point $\mathbf{s} = (s_0 s_1 s_2 \dots) \in \Sigma$. We are looking for a point x with $\iota(x) = \mathbf{s}$. Consider the set of $y \in \Lambda$ such that

$$d(\mathbf{s}, \iota(y)) \leq \frac{1}{2^n}.$$

This is the same as requiring that y belong to

$$\Lambda \cap I_{s_0 s_1 \dots s_n}$$

where $I_{s_0 s_1 \dots s_n}$ is the interval

$$I_{s_0 s_1 \dots s_n} = \{y \in I |\ y \in I_{s_0}, Q_c(y) \in I_{s_1}, \dots Q_c^{\circ n}(y) \in I_{s_n}\}.$$

So

$$\begin{aligned} I_{s_0 s_1 \dots s_n} &= I_{s_0} \cap Q_c^{-1}(I_{s_1}) \cap \dots \cap\ Q_c^{-n}(I_{s_n}) \\ &= I_{s_0} \cap Q_c^{-1}(I_{s_1} \cap \dots \cap\ Q_c^{-(n-1)}(I_{s_n})) \\ &= I_{s_0} \cap Q_c^{-1}(I_{s_1 \dots s_n}) \qquad (4.13) \\ &= I_{s_0 s_1 \dots s_{n-1}} \cap Q_c^{-n}(I_{s_n}) \subset I_{s_0 \dots s_{n-1}}. \qquad (4.14) \end{aligned}$$

The inverse image of any interval, J under Q_c consists of two intervals, one lying in I_0 and the other lying in I_1. For $n = 0$, I_{s_0} is either I_0 or I_1 and hence is an interval. By induction, it follows from (4.13) that $I_{s_0 s_1 \dots s_n}$ is an interval. By (4.14), these intervals are nested. By construction these nested intervals are closed. Since every sequence of closed nested intervals on the real line has a non-empty intersection, there is a point x which belongs to all of these intervals. Hence all the iterates of x lie in I, so $x \in \Lambda$and $\iota(x) = \mathbf{s}$. □

Continuity of ι.

The above argument shows that the interiors of the intervals $I_{s_0 s_1 \ldots s_n}$ (intersected withΛ) form neighborhoods of x that map into small neighborhoods of $\iota(x)$

Continuity of ι^{-1}.

Conversely, any small neighborhood of x in Λ will contain one of the intervals $I_{s_0 \ldots s_n}$ and hence all of the points $\mathbf{t}$ whose first n coordinates agree with $\mathbf{s} = \iota(x)$ will be mapped by ι^{-1} into the given neighborhood of x.

Summary.

To summarize: we have proved

Theorem 4.6.2. *Suppose that c satisfies (4.9). Let $\Lambda \subset [-p_+, p_+]$ consist of those points whose images under Q_c^n lie in $[-p,p_+]$ for all $n \geq 0$. Then Λ is a closed, non-empty, disconnected set. The itinerary map ι is a homeomorphism of Λ onto the sequence space, Σ, and conjugates Q_c to the shift map, S.*

Just as in the case of the space X above, the periodic points for S are precisely the periodic or "repeating" sequences. Thus we can conclude from the theorem that there are exactly 2^n points of period (at most) n for Q_c. Also, the same argument as above shows that the periodic points for S are dense inΣ, and hence the periodic points for Q_c are dense inΛ . Finally, the same argument as we gave in Section 4.4 shows that S is transitive onΣ . Hence, the restriction of Q_c to Λ is chaotic.

Chapter 5

Space and time averages.

5.1 Histograms and invariant densities.

5.1.1 Historgrams of iterations.

Let us consider a map, $F : [0,1] \to [0,1]$, pick an initial seed, x_0, and compute its iterates, $x_0, x_1, x_2, \ldots, x_m$ under F. We would like to see which parts of the unit interval are visited by these iterates, and how often. For this purpose let us divide the unit interval up into N subintervals of size $1/N$ given by

$$I_k = \left[\frac{k-1}{N}, \frac{k}{N}\right), \; k = 1, \ldots, N-1, \;\; I_N = \left[\frac{N-1}{N}, 1\right].$$

We count how many of the iterates $x_0, x_1, \ldots, x_m$ lie in I_k. Call this number n_k. There are $m+1$ iterates (starting with, and including, x_0) so the numbers

$$p_k = \frac{n_k}{m+1}$$

add up to one:

$$p_1 + \cdots + p_N = 1.$$

We would like to think of the p_k as "probabilities" - the number p_k representing the "probability" that an iterate belongs to I_k. Strictly speaking, we should write $p_k(m)$. In fact, we should write $p_k(m, x_0)$ since the procedure depends on the initial seed, x_0. But the hope is that as m gets large the $p_k(m)$ tend to a limiting value which we denote by p_k, and that this limiting value will be independent of x_0 if x_0 is chosen "generically".

We will continue in this vague, intuitive, vein a while longer before passing to a precise mathematical formulation. If U is a union of some of the I_k, then we can write

$$p(U) = \sum_{I_k \subset U} p_k$$

and think of $p(U)$ as representing the "probability" that an iterate of x_0 belongs to U. If N is large, so the intervals I_k are small, every open set U can be closely approximated by a union of the I_k's, so we can imagine that the "probabilities", $p(U)$, are defined for all open sets, U.

If we buy all of this, then we can write down an equation which has some chance of determining what these "probabilities", $p(U)$, actually are: A point $y = F(x)$ belongs to U if and only if $x \in F^{-1}(U)$. Thus the number of points among the $x_1, \ldots, x_{m+1}$ which belong to U is the same as the number of points among the $x_0, \ldots, x_m$ which belong to $F^{-1}(U)$. Since (we hope that) our limiting probability is unaffected by this shift from 0 to 1 or from m to $m+1$ we get the equation

$$p(U) = p(F^{-1}(U)). \tag{5.1}$$

To understand this equation, let us put it in a more general context. Suppose that we have a "measure", μ, which assigns a size, $\mu(A)$, to every open set, A. Let F be a continuous transformation. We then define the **push forward** measure, $F_*\mu$ by

$$(F_*\mu)(A) = \mu(F^{-1}(A)). \tag{5.2}$$

Without developing the language of measure theory, which is really necessary for a full understanding, we will try to describe some of the issues involved in the study of equations (5.2) and (5.1) from a more naive viewpoint. Consider, for example, $F = L_\mu, 1 < \mu < 3$. If we start with any initial seed other than $x_0 = 0$ or $x_0 = 1$, it is clear that the limiting probability is

$$p(I_k) = 1,$$

if the fixed point,$1 - \frac{1}{\mu} \in I_k$ and

$$p(I_k) = 0$$

otherwise.

Similarly, if $3 < \mu < 1 + \sqrt{6}$, and we start with any x_0 other than $0, 1, 1/\mu$, or the fixed point, $1 - \frac{1}{\mu}$ then clearly the limiting probability will be $p(I) = 1$ if both points of period two belong to I, $p(I) = \frac{1}{2}$ if I contains exactly one of the two period two points, and $p(I) = 0$ otherwise.

Discrete measures.

These are all examples of **discrete** measures in the sense that there is a finite (or countable) set of points, $\{z_k\}$, each assigned a positive number, $m(z_k)$ and

$$\mu(I) = \sum_{z_k \in I} m(z_k).$$

We are making the implicit assumption that this series converges for every bounded interval.

The "integral" with respect to a discrete measure.

The **integral** of a function, ϕ, with respect to the discrete measure, μ, denoted by $\langle \phi, \mu \rangle$ or by $\int \phi\mu$ is defined as

$$\int \phi\mu = \sum \phi(x_k)m(x_k).$$

This definition makes sense under the assumption that the series on the right hand side is absolutely convergent.

The push forward of a discrete measure.

The rule for computing the push forward, $F_*\mu$ (when defined) is very simple. Indeed, let $\{y_\ell\}$ be the set of points of the form $y_\ell = F(x_k)$ for some k, and set

$$n(y_\ell) = \sum_{F(x_k)=y_\ell} m(x_k).$$

Notice that there is some problem with this definition if there are infinitely many points x_k which map to the same y_l. Once again we must make some convergence assumption. For example, if the map F is everywhere finite-to-one, there will be no problem. Thus the push forward of a discrete measure is a discrete measure given by the above formula.

Absolutely continuous measures.

At the other extreme, a measure is called **absolutely continuous** (with respect to Lebesgue measure) if there is an integrable function, ρ, called the **density** so that

$$\mu(I) = \int_I \rho(x)dx.$$

For any continuous function, ϕ we define the integral of ϕ with respect to μ as

$$\langle \phi, \mu \rangle = \int \phi\mu = \int \phi(x)\rho(x)dx$$

if the integral is absolutely convergent.

Push forward of an absolutely continuous measure.

Suppose that the map F is piecewise differentiable and in fact satisfies $|F'(x)| \neq 0$ except at a finite number of points. These points are called *critical points* for the map F and their images are called *critical values.*

Suppose that A is an interval containing no critical values, and to fix the ideas, suppose that $F^{-1}(A)$ is the union of finitely many intervals, J_ℓ each of which is mapped monotonically (either strictly increasing or decreasing) onto

A. The change of variables formula from ordinary calculus says that for any function $g = g(y)$ we have

$$\int_A g(y)dy = \int_{J_k} g(F(x))|F'(x)|dx,$$

where $y = F(x)$.

So if we set $g(y) = \rho(x)|1/F'(x)|$ we get

$$\int \rho(x)\frac{1}{|F'(x)|}dy = \int_{J_k} \rho(x)dx = \mu(J_k).$$

Summing over k and using the definition (5.2) we see that $F_*\mu$ has the density (on any interval not containing a critical value) given by

$$\sigma(y) = \sum_{F(x_k)=y} \frac{\rho(x)}{|F'(x)|}. \tag{5.3}$$

Equation (5.3) is sometimes known as the Perron Frobenius equation, and the transformation $\rho \mapsto \sigma$ as the Perron Frobenius operator.

If $F : I \to I$ and μ has the density σ, then the equation

$$F_*\mu = \mu$$

requires that

$$\rho(y) = \sum_{F(x_k)=y} \frac{\rho(x)}{|F'(x)|}. \tag{5.4}$$

at regular values of F. We will see that this imposes severe and usable restrictions on ρ.

Back to the histogram

Getting back to our histogram, if we expect the limit measure to be of the absolutely continuous type, so

$$p(I_k) \approx \rho(x) \times \frac{1}{N}, \quad x \in I_k$$

then we expect that

$$\rho(x) \approx \lim_{m\to\infty} \frac{n_k N}{m+1}, \quad x \in I_k$$

as the formula for the limiting density.

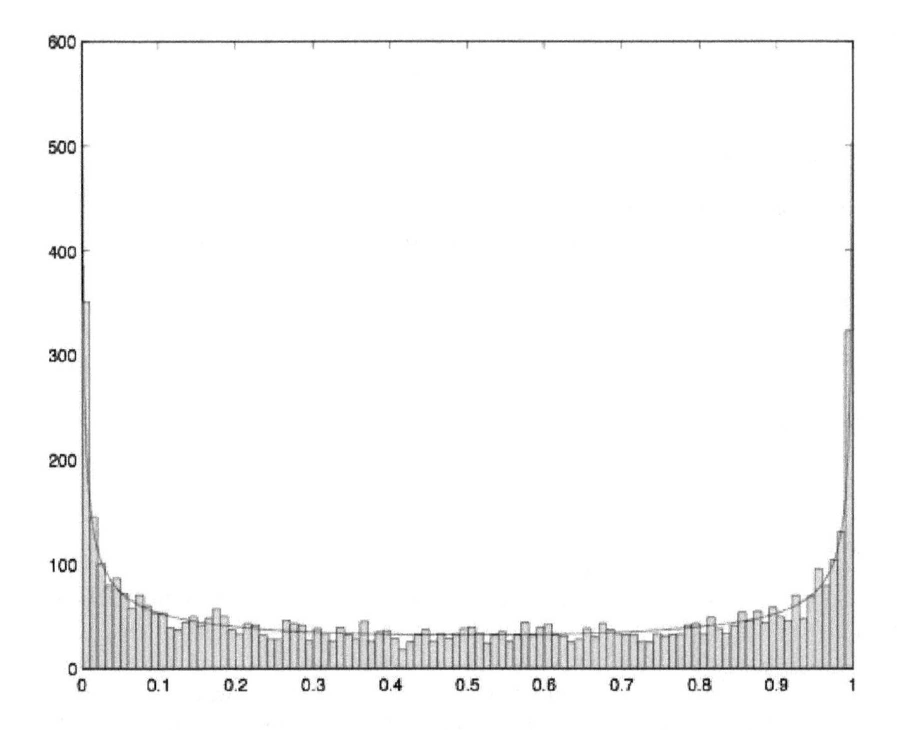

Figure 5.1: The histogram of L_4 compared with the function σ.

5.2 The histogram of L_4.

We wish to prove the following assertions:
(i) *The measure, μ, with density*

$$\sigma(x) = \frac{1}{\pi\sqrt{x(1-x)}} \tag{5.5}$$

is invariant under L_4. In other words it satisfies

$$L_{4*}\mu = \mu.$$

(ii) *Up to a multiplicative constant, (5.5) is the only continuous density invariant under L_4*
(iii) *If we pick the initial seed generically, then the normalized histogram converges to (5.5).* The m-file logistic4histogram.m which you can download from the website of this course will allow you to play with this result.

Figure 5.1 will give a typical result. We will give two proofs of (i). We wish to prove that

$$\rho(y) = \sum_{F(x_k)=y} \frac{\rho(x)}{|F'(x)|}. \tag{5.4}$$

where
$$\rho(x) = \frac{1}{\pi\sqrt{x(1-x)}}. \qquad (5.5)$$

First proof of (i).

The first is a direct verification of (5.4) with $y = F(x) = 4x(1-x)$ so $|F'(x)| = |F'(1-x)| = 4|1-2x|$. Notice that the ρ given by (5.5) satisfies $\sigma(x) = \sigma(1-x)$ so (5.4) becomes

$$\frac{1}{\pi\sqrt{4x(1-x)(1-4x(1-x))}} = \frac{2}{\pi 4|1-2x|\sqrt{x(1-x)}}.$$

This follows immediately from the identity

$$1 - 4x(1-x) = (2x-1)^2.$$

Second proof of (i).

The second proof is longer, but more instructive: Consider the tent transformation, T. For any interval, I contained in $[0,1]$, $T^{-1}(I)$ consists of the union of two intervals, each of half the length of I. In other words the ordinary Lebesgue measure is preserved by the tent transformation: $T_*\nu = \nu$ where ν has density $\rho(x) \equiv 1$. Put another way, the function $\rho(x) \equiv 1$ is the solution of the Perron Frobenius equation
$$\rho(Tx) = \frac{\rho(x)}{2} + \frac{\rho(1-x)}{2}. \qquad (5.6)$$
It follows immediately from the definitions, that

$$(F \circ G)_*\mu = F_*(G_*\mu), \qquad (5.7)$$

where F and G are two transformations , and μ is a measure.

In particular, since $h \circ T = L_4 \circ h$ where

$$h(x) = \sin^2 \frac{\pi x}{2},$$

it follows that if $T_*\nu = \nu$, then $L_{4*}(h_*\nu) = h_*\nu$. So to solve $L_{4*}\mu = \mu$, we must merely compute $h_*\nu$. According to (5.3) this is the measure with density

$$\sigma(y) = \frac{1}{|h'(x)|} = \frac{1}{\pi \sin\frac{\pi x}{2}\cos\frac{\pi x}{2}}.$$

But since $y = \sin^2\frac{\pi x}{2}$ this becomes

$$\sigma(y) = \frac{1}{\pi\sqrt{y(1-y)}}$$

as desired.

Proof of (ii).

To prove (ii), it is enough to prove the corresponding result for the tent transformation: that $\rho =$ const. is the only continuous function satisfying (5.6). To prove this assertion about T, this, let us consider the binary representation of T:

Let
$$x = 0.a_1a_2a_3\ldots$$
be the binary expansion of x. If $0 \le x < \frac{1}{2}$, so $a_1 = 0$, then $Tx = 2x$ or
$$T(0.0a_2a_3a_4\ldots) = 0.a_2a_3a_4\ldots.$$
If $x \ge \frac{1}{2}$, so $a_1 = 1$, then
$$T(x) = -2x + 2 = 1 - (2x - 1) = 1 - S(x) = 1 - 0.a_2a_3a_4\ldots.$$
Introducing the notation
$$\bar{0} = 1, \quad \bar{1} = 0,$$
we have
$$0.a_2a_3a_4\cdots + 0.\bar{a}_1\bar{a}_2\bar{a}_3\cdots = 0.1111\cdots = 1$$
so
$$T(0.1a_2a_3a_4\ldots) = 0.\bar{a}_2\bar{a}_3\bar{a}_4\ldots.$$
In particular, $T^{-1}(0.a_1a_2a_3\ldots)$ consists of the two points
$$0.0a_1a_2a_3\ldots \quad \text{and} \quad 0.1\bar{a}_1\bar{a}_2\bar{a}_3\ldots.$$
Now let us iterate (5.6) with ρ replaced by f, and rewrite this equation as
$$f(x) = \frac{1}{2}[f(u) + f(v)]$$
where $Tu = Tv = x$. We want to show that the only continuous solution of this equation is is $f =$ constant. Using the notation
$$x = 0.a_1a_2\cdots = 0.a,$$
a repeated application of (5.6) gives:
$$\begin{aligned}
f(x) &= \frac{1}{2}[f(.0a) + f(.1\bar{a})] \\
&= \frac{1}{4}[f(.00a) + f(.01\bar{a}) + f(.10a) + f(.11\bar{a})] \\
&= \frac{1}{8}[f(.000a) + f(.001\bar{a}) + f(.010a) + f(.011\bar{a}) + f(.100a) + \cdots] \\
&\to \int f(t)dt.
\end{aligned}$$
But this integral is a constant, independent of x. □

What about (iii)?

The third statement, (iii), about the limiting histogram for "generic" initial seed, x_0, demands a more careful formulation. What do we mean by the phrase "generic"? The precise formulation requires a dose of measure theory: the word "generic" should be taken to mean "outside of a set of measure zero with respect to μ". The usual phrase for this is "for almost all x_0". Then assertion (iii) becomes a special case of the famous Birkhoff ergodic theorem.

In our case, the Birkhoff ergodic theorem asserts that for almost all points, p, the "time average"

$$\lim \frac{1}{n}\sum_{k=0}^{n-1}\phi(L_4^k p)$$

equals the "space average"

$$\int \phi\mu$$

for any integrable function, ϕ. Rather than proving this theorem, I will explain a simpler theorem, von Neumann's mean ergodic theorem, which motivated Birkhoff to prove his theorem.

5.3 The mean ergodic theorem.

Let F be a transformation with an invariant measure, μ. By this we mean that $F_*\mu = \mu$. We let H denote the Hilbert space of all square integrable functions with respect to μ, so the scalar product of $f, g \in H$ is given by

$$(f,g) = \int f\bar{g}\mu.$$

The map F induces a transformation $U : H \to H$ by

$$Uf = f \circ F$$

and

$$(Uf, Ug) = \int (f\circ F)(\overline{g\circ F})\mu = \int f\bar{g}\mu = (f,g).$$

In other words, U is an isometry of H. The mean ergodic theorem asserts that the limit of

$$\frac{1}{n}\sum_{0}^{n-1} U^k f$$

exists in the Hilbert space sense, "convergence in mean", rather than the almost everywhere pointwise convergence of the Birkhoff ergodic theorem. Practically by its definition, this limiting element $\hat{f}$ is invariant, i.e. satisfies $U\hat{f} = \hat{f}$. Indeed, applying U to the above sum gives an expression which differs from that sum by only two terms, f and $U^n f$ and dividing by n sends these terms to zero as $n \to \infty$. If, as in our example, we know what the possible invariant elements are, then we know the possible limiting values $\hat{f}$

The mean ergodic theorem can be regarded as a smeared out version of the Birkhoff theorem. Due to inevitable computer error, the mean ergodic theorem may actually be the version that we want.

So we wish to prove:

Theorem 5.3.1. **[von Neumann's mean ergodic theorem.]** *Let $U : H \to H$ be an isometry of a Hilbert space, H. Then for any $f \in H$, the limit*

$$\lim \frac{1}{n} \sum U^k f = \hat{f} \tag{5.8}$$

exists in the Hilbert space sense, and the limiting element $\hat{f}$ is invariant, i.e. $U\hat{f} = \hat{f}$.

The limit, if it exists, is invariant as we have seen. If U were a unitary operator on a finite dimensional Hilbert space, H, then we could diagonalize U, and hence reduce the theorem to the one dimensional case. A unitary operator on a one dimensional space is just multiplication by a complex number of the form $e^{i\alpha}$. If $e^{i\alpha} \neq 1$, then

$$\frac{1}{n}(1 + e^{i\alpha} + \cdots + e^{(n-1)i\alpha}) = \frac{1}{n} \frac{1 - e^{in\alpha}}{1 - e^{i\alpha}} \to 0.$$

On the other hand, if $e^{i\alpha} = 1$, the expression on the left is identically one. This proves the theorem for finite dimensional unitary operators.

Proof in general. For an infinite dimensional Hilbert space, we could apply the spectral theorem of Stone (discovered shortly before the proof of the ergodic theorem) and this was von Neumann's original method of proof.

Actually, we can give the following proof due to F. Riesz:

Lemma 5.3.1. *The orthogonal complement of the set, D, of all elements of the form $Ug - g$, consists of invariant elements.*

Proof of the lemma. If f is orthogonal to all elements in D, then,in particular, f is orthogonal to $Uf - f$, so

$$0 = (f, Uf - f)$$

and

$$(Uf, Uf - f) = (Uf, Uf) - (Uf, f) = (f, f) - (Uf, f)$$

since U is an isometry. So

$$(Uf, Uf - f) = (f - Uf, f) = 0.$$

So

$$(Uf - f, Uf - f) = (Uf, Uf - f) - (f, Uf - f) = 0,$$

or

$$Uf - f = 0$$

which says that f is invariant. □ So what we have shown, in fact, is

Lemma 5.3.2. *The union of the set D with the set, I, of the invariant functions is dense in H.*

Indeed, if f is orthogonal to D, then it must be invariant, and if it is orthogonal to all invariant functions it must be orthogonal to itself, and so must be zero. So $(D \cup I)^\perp = 0$, so $D \cup I$ is dense in H.

If f is invariant, then clearly the limit(5.8) exists and equals f. If $f = Ug - g$, then the expression on the left in (5.8) telescopes into

$$\frac{1}{n}(U^n g - g)$$

which clearly tends to zero. Hence, as a corollary we obtain

Lemma 5.3.3. *The set of elements for which the limit in (5.8) exists is dense in H.*

Completion of the proof of the mean ergodic theorem. Hence the mean ergodic theorem will be proved, once we prove

Lemma 5.3.4. *The set of elements for which the limit in (5.8) exists is closed.*

Proof.

$$\frac{1}{n}\sum U^k g_i \to \hat{g}_i, \quad \frac{1}{n}\sum U^k g_j \to \hat{g}_j,$$

and

$$\| g_i - g_j \| < \epsilon,$$

then

$$\| \frac{1}{n}\sum U^k g_i - \frac{1}{n}\sum U^k g_j \| < \epsilon,$$

so

$$\| \hat{g}_i - \hat{g}_j \| < \epsilon.$$

So if $\{g_i\}$ is a sequence of elements converging to f, we conclude that $\{\hat{g}_i\}$ converges to some element, call it $\hat{f}$. If we choose i sufficiently large so that $\| g_i - f \| < \epsilon$, then

$$\| \frac{1}{n}\sum U^k f - \hat{f} \| \leq \| \frac{1}{n}\sum U^k (f - g_i) \| + \| \frac{1}{n}\sum U^k g_i - \hat{g}_i \| + \| \hat{g}_i - \hat{f} \| \leq 3\epsilon,$$

proving the lemma and hence proving the mean ergodic theorem. □

5.4 The arc sine law.

The probability distribution with density

$$\sigma(x) = \frac{1}{\pi\sqrt{x(1-x)}}$$

is called the **arc sine law** in probability theory because, if I is the interval $I = [0, u]$ then

$$\text{Prob } x \in I = \text{Prob } 0 \leq x \leq u = \int_0^u \frac{1}{\pi\sqrt{x(1-x)}} = \frac{2}{\pi} \arcsin \sqrt{u}. \tag{5.9}$$

We have already verified this integration because $I = h(J)$ where

$$h(t) = \sin^2 \frac{\pi t}{2}, \quad J = [0, v], \ h(v) = u,$$

and the probability measure we are studying is the push forward of the uniform distribution. So

$$\text{Prob } h(t) \in I = \text{Prob } t \in J = v.$$

We could, of course, verify the integration directly.

The arc sine law plays a crucial role in the theory of fluctuations in random walks. As a cultural diversion we explain some of the key ideas, following the treatment in [Feller] very closely.

5.4.1 The random walk.

Suppose that there is an ideal coin tossing game in which each player wins or loses a unit amount with (independent) probability $\frac{1}{2}$ at each throw. Let $S_0 = 0, S_1, S_2, \ldots$ denote the successive cumulative gains (or losses) of the first player. We can think of the values of these cumulative gains as being marked off on a vertical s-axis, and representing the position of a particle which moves up or down with probability $\frac{1}{2}$ at each (discrete) time unit .

Let

$$\alpha_{2k,2n}$$

denote the *probability that up to and including time* $2n$*, the last visit to the origin occurred at time* $2k$. Let

$$u_{2\nu} = \binom{2\nu}{\nu} 2^{-2\nu}. \tag{5.10}$$

So $u_{2\nu}$ represents the probability that exactly ν out of the first 2ν steps were in the positive direction, and the rest in the negative direction. In other words, $u_{2\nu}$ is the probability that the particle has returned to the origin at time 2ν.

Using Stirling's formula.

We can find a simple approximation to $u_{2\nu}$ using Stirling's formula for an approximation to the factorial:

$$n! \sim \sqrt{2\pi} n^{n+\frac{1}{2}} e^{-n}$$

where the $\sim$ signifies that the ratio of the two sides tends to one as n tends to infinity.

There are many proofs of Stirling's formula. I will give two at the end of this section. In the meanwhile, let's take it for granted.

Then

$$\begin{aligned} u_{2\nu} &= 2^{-2\nu}\frac{(2\nu)!}{(\nu!)^2} \\ &\sim 2^{-2\nu}\frac{\sqrt{2\pi}(2\nu)^{2\nu+\frac{1}{2}}e^{-2\nu}}{2\pi\nu^{2\nu+1}e^{-2\nu}} \\ &= \frac{1}{\sqrt{\pi\nu}}. \end{aligned}$$

The results we wish to prove in this section are:

Proposition 5.4.1. *We have*

$$\alpha_{2k,2n} = u_{2k}u_{2n-2k}, \tag{5.11}$$

so we have the asymptotic approximation

$$\alpha_{2k,2n} \sim \frac{1}{\pi\sqrt{k(n-k)}}. \tag{5.12}$$

If we set

$$x_k = \frac{k}{n}$$

then we can write

$$\alpha_{2k,2n} \sim \frac{1}{n}\sigma(x_k). \tag{5.13}$$

Thus, for fixed $0 < x < 1$ and n sufficiently large

$$\sum_{k<xn} \alpha_{2k,2n} \doteq \frac{2}{\pi}\arcsin\sqrt{x}. \tag{5.14}$$

Proposition 5.4.2. *The probability that in the time interval from 0 to $2n$ the particle spends $2k$ time units on the positive side and $2n-2k$ time units on the negative side equals $\alpha_{2k,2n}$. In particular, if $0 < x < 1$ the probability that the fraction k/n of time units spent on the positive be less than x tends to $\frac{2}{\pi}\arcsin\sqrt{x}$ as $n \to \infty$.*

Let us call the value of S_{2n} for any given realization of the random walk, the **terminal point**. Of course, the particle may well have visited this terminal point earlier in the walk, and we can ask when it first reaches its terminal point.

Proposition 5.4.3. *The probability that the first visit to the terminal point occurs at time* $2k$ *is given by* $\alpha_{2k,2n}$.

We can also ask for the first time that the particle reaches its maximum value: We say that the **first maximum occurs at time** l if

$$S_0 < S_l, S_1 < S_l, \ldots S_{l-1} < S_l, \qquad S_{l+1} \leq S_l, S_{l+2} \leq S_l, \ldots S_{2n} \leq S_l. \quad (5.15)$$

Proposition 5.4.4. *If* $0 < l < 2n$ *the probability that the first maximum occurs at* $l = 2k$ *or* $l = 2k+1$ *is given by* $\frac{1}{2}\alpha_{2k,2n}$. *For* $l = 0$ *this probability is given by* u_{2n} *and if* $l = 2n$ *it is given by* $\frac{1}{2}u_{2n}$.

Before proving these various facts, let us discuss a few of their implications which some people find counterintuitive. For example, because of the shape of the density, σ, the last result implies that the maximal accumulated gain is much more likely to occur very near to the beginning or to the end of a coin tossing game rather than somewhere in the middle. The fourth assertion implies that the probability that the first visit to the terminal point occurs at time $2k$ is that same as the probability that it occurs at time $2n - 2k$ and that very early first visits and very late first visits are much more probable than first visits some time in the middle.

In order to get a better feeling for the assertion of the first two propositions, let us tabulate the values of $\frac{2}{\pi}\arcsin\sqrt{x}$ for $0 \leq x \leq \frac{1}{2}$.

x	$\frac{2}{\pi}\arcsin\sqrt{x}$	x	$\frac{2}{\pi}\arcsin\sqrt{x}$
0.05	0.144	0.30	0.369
0.10	0.205	0.35	0.403
0.15	0.253	0.40	0.236
0.20	0.295	0.45	0.468
0.25	0.333	0.50	0.500

This table, in conjunction with our Propositions 5.4.1 and 5.4.2 says that if a great many coin tossing games are conducted every second, day and night for a hundred days, then in about 14.4 percent of the cases,the lead will not change after day five.

5.4.2 The reflection principle.

The proof of all four propositions hinges on three lemmas. Let us graph (by a polygonal path) the walk of a particle. So a "path" is a broken line segment made up of segments of slope ± 1 joining integral points to integral points in the plane (with the time or $t-$axis horizontal and the $s-$axis vertical). If $A = (a,\alpha)$ is a point, we let $A' = (a, -\alpha)$ denote its image under reflection in the $t-$axis.

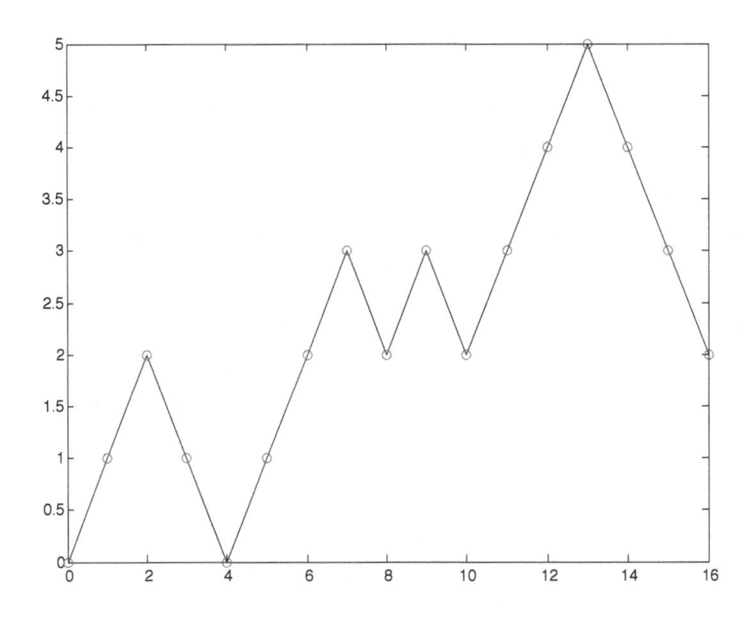

Figure 5.2: Original path.

Lemma 5.4.1. The reflection principle. *Let $A = (a,\alpha), B = (b,\beta)$ be points in the first quadrant with $b > a \geq 0, \alpha > 0, \beta > 0$. The number of paths from A to B which touch or cross the $t-$ axis equals the number of all paths from A' to B.*

Proof. For any path from A to B which touches the horizontal axis, let t be the abscissa of the first point of contact. Reflect the portion of the path from A to $T = (t, 0)$ relative to the horizontal axis. This reflected portion is a path from A' to T, and continues to give a path from A' to B. This procedure assigns to each path from A to B which touches the axis, a path from A' to B. This assignment is bijective: Any path from A' to B must cross the $t-$axis. Reflect the portion up to the first crossing to get a touching path from A to B. This is the inverse assignment. □

A path with n steps will join $(0,0)$ to (n,x) if and only if it has p steps of slope $+1$ and q steps of slope -1 where

$$p+q=n, \quad p-q=x.$$

The number of such paths is the number of ways of picking the positions of the p steps of positive slope and so the number of paths joining $(0,0)$ to (n,x) is

$$N_{n,x} = \binom{p+q}{p} = \binom{n}{\frac{n+x}{2}}.$$

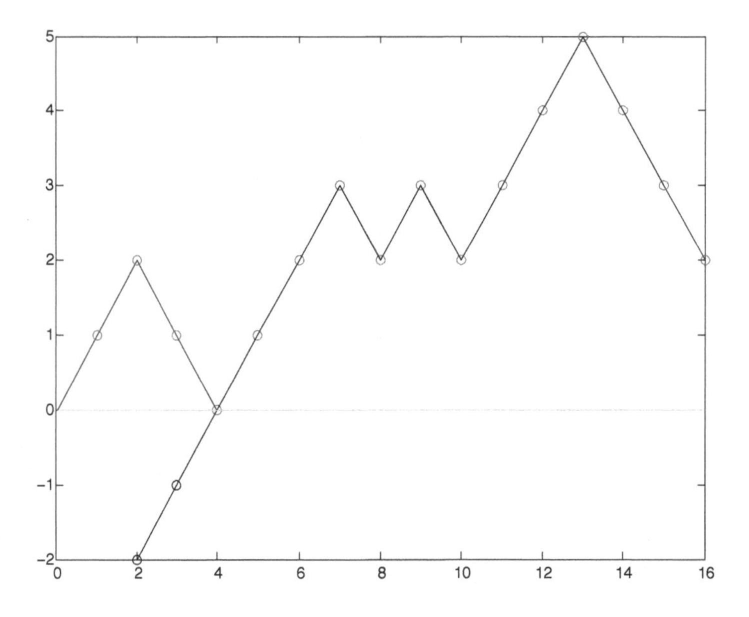

Figure 5.3: Reflected path.

It is understood that this formula means that $N_{n,x} = 0$ when there are no paths joining the origin to (n, x).

Lemma 5.4.2. The ballot theorem. *Let n and x be positive integers. There are exactly*

$$\frac{x}{n} N_{n,x}$$

paths which lie strictly above the t axis for $t > 0$ and join $(0,0)$ to (n, x).

Proof. There are as many paths joining $(0,0)$ to (n, x) which are strictly above the x-axis as there are paths joining $(1,1)$ to (n, x) which do *not* touch or cross the $t-$axis. This is the same as the total number of paths which join $(1,1)$ to (n, x) less the number of paths which do touch or cross. By the reflection principle, the number of paths which do touch or cross is the same as the number of paths joining $(1, -1)$ to (n, x) which is $N_{n-1,x+1}$. Thus, with p and q as above, the number of paths which lie strictly above the t axis for $t > 0$ and which join

$(0,0)$ to (n,x) is

$$\begin{aligned} N_{n-1,x-1} - N_{n-1,x+1} &= \binom{p+q-1}{p-1} - \binom{p+q-1}{p} \\ &= \frac{(p+q-1)!}{(p-1)!(q-1)!}\left[\frac{1}{q} - \frac{1}{p}\right] \\ &= \frac{p-q}{p+q} \times \frac{(p+q)!}{p!q!} \\ &= \frac{x}{n} N_{n,x} \quad \square \end{aligned}$$

□

The reason that this lemma is called the Ballot Theorem is that it asserts that if candidate P gets p votes, and candidate Q gets q votes in an election where the probability of each vote is independently $\frac{1}{2}$, and if P wins, i.e. if $p > q$, then the probability that throughout the counting there are more votes for P than for Q is given by

$$\frac{p-q}{p+q}.$$

Here is our last lemma:

Lemma 5.4.3. *The probability that from time* 1 *to time* $2n$ *the particle stays strictly positive is given by* $\frac{1}{2}u_{2n}$. *In symbols,*

$$\text{Prob } \{S_1 > 0, \dots, S_{2n} > 0\} = \frac{1}{2}u_{2n}. \tag{5.16}$$

So

$$\text{Prob } \{S_1 \neq 0, \dots, S_{2n} \neq 0\} = u_{2n}. \tag{5.17}$$

Also

$$\text{Prob } \{S_1 \geq 0, \dots, S_{2n} \geq 0\} = u_{2n}. \tag{5.18}$$

Proof. By considering the possible positive values of S_{2n} which can range from 2 to $2n$ we have Prob $\{S_1 > 0, \dots, S_{2n} > 0\}$

$$\begin{aligned} &= \sum_{r=1}^{n} \text{Prob } \{S_1 > 0, \dots, S_{2n} = 2r\} \\ &= 2^{-2n} \sum_{r=1}^{n} (N_{2n-1,2r-1} - N_{2n-1,2r+1}) \\ &= 2^{-2n} (N_{2n-1,1} - N_{2n-1,3} + N_{2n-1,3} - N_{2n-1,5} + \cdots) \\ &= 2^{-2n} N_{2n-1,1}. \end{aligned}$$

The passage from the first line to the second is the reflection principle, as in our proof of the Ballot Theorem, from the third to the fourth is because the sum telescopes to $N_{2n-1,1} - N_{2n-1,2n+1}$ and $N_{2n-1,2n+1} = 0$ because you can't get from 0 to $2n+1$ in $2n-1$ steps.

So

$$\begin{aligned}\text{Prob}\,\{S_1 > 0, \ldots, S_{2n} > 0\} &= 2^{-2n} N_{2n-1,1} \\ &= \frac{1}{2} p_{2n-1,1} \\ &= \frac{1}{2} u_{2n}.\end{aligned}$$

The $p_{2n-1,1}$ on the next to the last line is the probability of ending up at $(2n-1, 1)$ starting from $(0, 0)$. The last equality is simply the assertion that to reach zero at time $2n$ we must be at ± 1 at time $2n-1$ (each of these has equal probability, $p_{2n-1,1}$) and for each alternative there is a 50 percent chance of getting to zero on the next step. This proves (5.16).

Since a path which never touches the $t-$axis must be always above or always below the $t-$axis, (5.17) follows immediately from (5.16).

Observe that a path which is strictly above the axis from time 1 on, must pass through the point $(1, 1)$ and then stay above the horizontal line $s = 1$. The probability of going to the point $(1, 1)$ at the first step is $\frac{1}{2}$, and then the probability of remaining above the new horizontal axis is Prob $\{S_1 \geq 0, \ldots, S_{2n-1} \geq 0\}$. But since $2n-1$ is odd, if $S_{2n-1} \geq 0$ then $S_{2n} \geq 0$. So, by (5.16) we have

$$\begin{aligned}\frac{1}{2} u_{2n} &= \text{Prob}\,\{S_1 > 0, \ldots, S_{2n} > 0\} \\ &= \frac{1}{2}\text{Prob}\,\{S_1 \geq 0, \ldots, S_{2n-1} \geq 0\} \\ &= \frac{1}{2}\text{Prob}\,\{S_1 \geq 0, \ldots, S_{2n-1} \geq 0, S_{2n} \geq 0\},\end{aligned}$$

completing the proof of the lemma. □

Proofs of the propositions.

Proof of Prop. 5.4.1. To say that the last visit to the origin occurred at time $2k$ means that

$$S_{2k} = 0$$

and

$$S_j \neq 0, \quad j = 2k+1, \ldots, 2n.$$

Recall that

$$u_{2\nu} = \binom{2\nu}{\nu} 2^{-2\nu}$$

is the probability that the particle has returned to the origin at time 2ν.

By definition, the first $2k$ positions can be chosen in $2^{2k} u_{2k}$ ways to satisfy the condition $S_{2k} = 0$. Taking the point $(2k, 0)$ as our new origin, (5.17) says that there are $2^{2n-2k} u_{2n-2k}$ ways of choosing the last $2n-2k$ steps so as to satisfy the condition $S_j \neq 0, \quad j = 2k+1, \ldots, 2n$. Multiplying and then dividing the result by 2^{2n} proves (5.11). □

Proof of Prop. 5.4.2. We consider paths of $2n$ steps and let $b_{2k,2n}$ denote the probability that exactly $2k$ sides lie above the t–axis. We want to show that

$$b_{2k,2n} = \alpha_{2k,2n}.$$

For the case $k = n$ we have $\alpha_{2n,2n} = u_0 u_{2n} = u_{2n}$ and $b_{2n,2n}$ is the probability that the path lies entirely above the axis. So our assertion reduces to (5.18) which we have already proved. By symmetry, the probability of the path lying entirely below the the axis is the same as the probability of the path lying entirely above it, so $b_{0,2n} = \alpha_{0,2n}$ as well.

So we need prove our assertion for $1 \leq k \leq n-1$. In this situation, a return to the origin must occur. Suppose that the first return to the origin occurs at time $2r$. There are then two possibilities: the entire path from the origin to $(2r, 0)$ is either above the axis or below the axis. If it is above the axis, then $r \leq k \leq n-1$, and the section of the path beyond $(2r, 0)$ has $2k - 2r$ edges above the t–axis. The number of such paths is

$$\frac{1}{2} 2^{2r} f_{2r} 2^{2n-2r} b_{2k-2r,2n-2r}$$

where f_{2r} denotes the *probability of first return* at time $2r$:

$$f_{2r} = \text{Prob}\,\{S_1 \neq 0, \ldots, S_{2r-1} \neq 0, S_{2r} = 0\}.$$

If the first portion of the path up to $2r$ is spent below the axis, the the remaining path has exactly $2k$ edges above the axis, so $n - r \geq k$ and the number of such paths is

$$\frac{1}{2} 2^{2r} f_{2r} 2^{2n-2r} b_{2k,2n-2r}.$$

So we get the recursion relation

$$b_{2k,2n} = \frac{1}{2} \sum_{r=1}^{k} f_{2r} b_{2k-2r,2n-2r} + \frac{1}{2} \sum_{r=1}^{n-k} f_{2r} b_{2k,2n-2r} \quad 1 \leq k \leq n-1. \tag{5.19}$$

Now we proceed by induction on n. We know that $b_{2k,2n} = u_{2k} u_{2n-2k} = \frac{1}{2}$ when $n = 1$. Assuming the result up through $n-1$, the recursion formula (5.19) becomes

$$b_{2k,2n} = \frac{1}{2} u_{2n-2k} \sum_{r=1}^{k} f_{2r} u_{2k-2r} + \frac{1}{2} u_{2k} \sum_{r=1}^{n-k} f_{2r} u_{2n-2k-2r}. \tag{5.20}$$

We claim that the probabilities of return and the probabilities of first return are related by

$$u_{2n} = f_2 u_{2n-2} + f_4 u_{2n-4} + \cdots + f_{2n} u_0. \tag{5.21}$$

Indeed, if a return occurs at time $2n$, then there must be a first return at some time $2r \leq 2n$ and then a return in $2n-2r$ units of time, and the sum in (5.21) is over the possible times of first return. If we substitute (5.21) into the first sum in (5.20) it becomes u_{2k} while substituting (5.21) into the second term yields u_{2n-2k}. Thus (5.20) becomes

$$b_{2k,2n} = u_{2k}u_{2n-2k}$$

which is our desired result. □

Proof of Prop. 5.4.3. The probability in the Proposition is the probability that $S_{2k} = S_{2n}$ but $S_j \neq S_{2n}$ for $j < 2k$. Reading the path rotated through 180° about the end point, and with the endpoint shifted to the origin, this is clearly the same as the probability that $2n - 2k$ is the last visit to the origin. □

Proof of Prop. 5.4.3. The probability that the maximum is achieved at 0 is the probability that $S_1 \leq 0, \dots, S_{2n} \leq 0$ which is u_{2n} by (5.18). The probability that the maximum is first obtained at the terminal point, is, after rotation and translation, the same as the probability that $S_1 > 0, \dots, S_{2n} > 0$ which is $\frac{1}{2}u_{2n}$ by (5.16). If the maximum occurs first at some time l in the middle, we combine these results for the two portions of the path - before and after time ℓ - together with (5.11) to complete the proof. □

A computer generated illustration..

Figure 5.4 shows a computer generated random walk with 100,000 steps. The last zero is at time 3783. For the remaining 96,217 steps the path is positive. According to the arc sine law, with probability 1/5, the particle will spend about 97.6 percent of its time on one side of the origin.

5.5 The Beta distributions.

The arc sine law is the special case $a = b = \frac{1}{2}$ of the Beta distribution with parameters a, b which has probability density proportional to

$$t^{a-1}(1-t)^{b-1}.$$

So long as $a > 0$ and $b > 0$ the integral

$$B(a,b) = \int_0^1 t^{a-1}(1-t)^{b-1}dt$$

converges, and was evaluated by Euler to be

$$B(a,b) = \frac{\Gamma(a)\Gamma(b)}{\Gamma(a+b)}$$

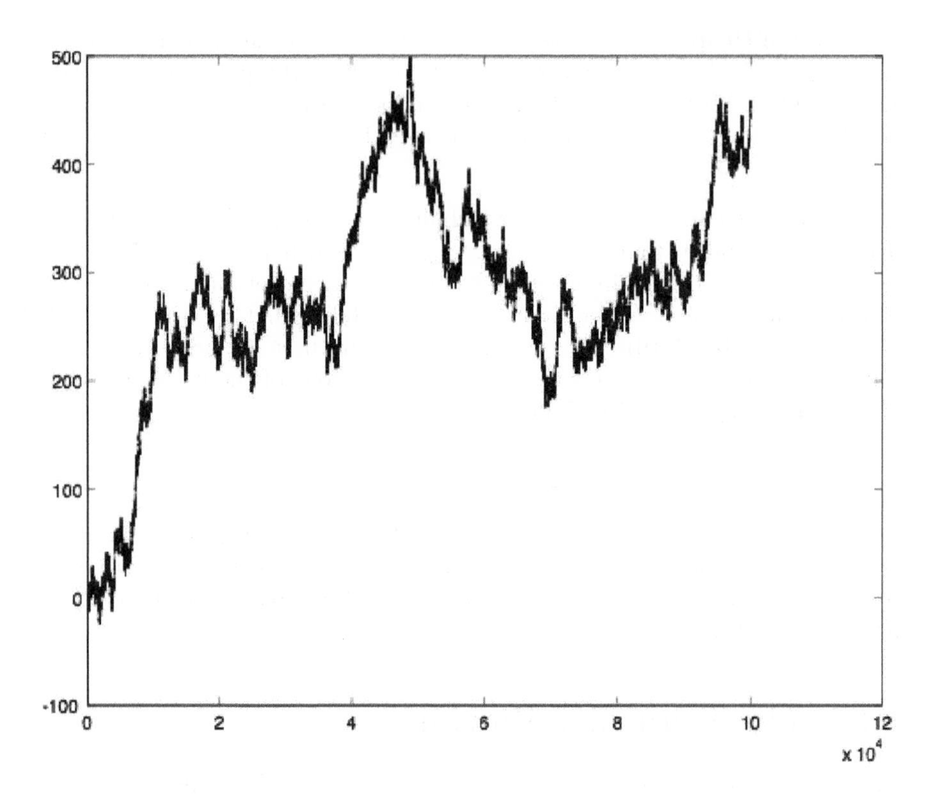

Figure 5.4: A random walk with 100,000 steps.

where Γ is Euler's Gamma function. So the Beta distributions with $a > 0, b > 0$ are given by

$$\frac{1}{B(a,b)} t^{a-1}(1-t)^{b-1}.$$

We characterized the arc sine law ($a = b = \frac{1}{2}$) as being the unique probability density invariant under L_4. The case $a = b = 0$, where the integral does not converge, also has an interesting characterization as an invariant density:

Consider transformations of the form

$$t \mapsto \frac{at+b}{ct+d}$$

where the matrix

$$\begin{pmatrix} a & b \\ c & d \end{pmatrix}$$

is invertible. Suppose we require that the transformation preserve the origin and the point $t = 1$. Preserving the origin requires that $b = 0$, while preserving the point $t = 1$ requires that $a = c + d$. Since $b = 0$ we must have $ad \neq 0$ for the matrix to be invertible. Since multiplying all the entries of the matrix by the same non-zero scalar does not change the transformation, we may as well assume that $d = 1$, and hence the family transformations we are looking at are

$$\phi_a: \quad t \mapsto \frac{at}{(a-1)t+1}, \qquad a \neq 0.$$

Notice that

$$\phi_a \circ \phi_b = \phi_{ab}.$$

Our claim is that, up to scalar multiple, the density

$$\rho(t) = \frac{1}{t(1-t)}$$

is the unique density such that the measure

$$\rho(t)dt$$

is invariant under all the transformations ϕ_a.

Indeed,

$$\phi_a'(t) = \frac{a}{[1-t+at]^2}$$

so the condition of invariance is

$$\frac{a}{[1-t+at]^2}\rho(\phi_a(t)) = \rho(t).$$

Let us normalize ρ by

$$\rho\left(\frac{1}{2}\right) = 4.$$

Then

$$s = \phi_a\left(\frac{1}{2}\right) \Leftrightarrow s = \frac{a}{1+a} \Leftrightarrow a = \frac{s}{1-s}.$$

So taking $t = \frac{1}{2}$ in the condition for invariance and a as above, we get

$$\rho(s) = 4((1-s)/s)[\frac{1}{2} + \frac{1}{2}\frac{s}{1-s}]^2 = \frac{1}{s(1-s)}.$$

This elementary geometrical fact - that $1/t(1-t)$ is the unique density (up to scalar multiple) which is invariant under all the ϕ_a - was given a deep philosophical interpretation by Jaynes, [Jaynes]: Suppose we have a possible event which may or may not occur, and we have a population of individuals each of whom has a clear opinion (based on ingrained prejudice, say, from reading the newspapers or watching television) of the probability of the event being true. So Mr. A assigns probability p(A) to the event E being true and (1-p(A)) as the probability of its not being true, while Mr. B assigns probability P(B) to its being true and (1-p(B)) to its not being true and so on.

Suppose an additional piece of information comes in, which would have a (conditional) probability x of being generated if E were true and y of this information being generated if E were not true. We assume that both x and y are positive, and that every individual thinks rationally in the sense that on the advent of this new information he changes his probability estimate in accordance with Bayes' law, which says that the posterior probability p' is given in terms of the prior probability p by

$$p' = \frac{px}{px + (1-p)y} = \phi_a(p) \quad \text{where} \quad a := \frac{x}{y}.$$

We might say that the population as a whole has been "invariantly prejudiced" if any such additional evidence does not change the proportion of people within the population whose belief lies in a small interval. Then the density describing this state of knowledge (or rather of ignorance) must be the density

$$\rho(p) = \frac{1}{p(1-p)}.$$

According to this reasoning of Jaynes, we take the above density to describe the prior probability an individual (thought of as a population of subprocessors in his brain) would assign to the probability of an outcome of a given experiment. If a series of experiments then yielded M successes and N failures, Bayes' theorem (in its continuous version) would then yield the posterior distribution of probability assignments as being proportional to

$$p^{M-1}(1-p)^{N-1}$$

the Beta distribution with parameters M, N.

5.6 Two proofs of Stirling's formula

I am going to give two rather different looking proofs Stirling's formula. Both illustrate important methods in asymptotic analysis. A more elementary proof can be found in [Feller]

5.6.1 The Euler-Maclauren summation formula.

Let f be a continuously differentiable function on $[0, n]$. Integration by parts shows that for every integer $k = 0, 1, \dots, n-1$ we have

$$\int_k^{k+1} [x - k - \frac{1}{2}] f'(x) dx = \frac{1}{2}[f_k + f_{k+1}] - \int_k^{k+1} f(x) dx$$

where we have written f_k for $f(k)$ to shorten the formula.

Letting $[x]$ denote the largest integer $\leq x$ we can write this as

$$\frac{1}{2}[f_k + f_{k+1}] = \int_k^{k+1} f(x) dx + \int_k^{k+1} [x - [x] - \frac{1}{2}] f'(x) dx.$$

Summing this from 0 to $n-1$ and adding $\frac{1}{2}[f_0 + f_n]$ gives (a baby version of) the **Euler-Maclauren formula**

$$f_0 + \cdots + f_n = \frac{1}{2}[f_0 + f_n] + \int_0^n f(x) dx + \int_0^n P_1(x) f'(x) dx$$

where

$$P_1(x) := x - [x] - \frac{1}{2}.$$

The function $P_1(x)$ is periodic of period one and is continuous except at the integers, and its integral over any interval of length one is zero. We will let P_2 be an indefinite integral of P_1. To be specific, let us define

$$P_2(x) := \frac{1}{2}x^2 - \frac{1}{2}x + \frac{1}{12} \quad 0 \leq x \leq 1$$

and extended to be periodic of period one. Its integral over any interval of length one is then also zero.

Deriving Stirling's formula from the Euler-Maclauren summation formula.

Let us apply the Euler-Maclauren summation formula to the function $f(x) = \frac{1}{1+x}$ and replace n by $n-1$ in the formula. We get

$$\log 1 + \log 2 + \cdot + \log n = \int_1^n \log x dx + \frac{1}{2} \log n + \int_1^n \frac{P_1(x)}{x} dx$$

$$= (n + \frac{1}{2}) \log n - (n-1) + \int_1^n \frac{P_1(x)}{x} dx.$$

Integration by parts gives

$$\int_1^n \frac{P_1(x)}{x} dx = \left. \frac{P_2(x)}{x} \right|_1^n + \int_1^n \frac{P_2(x)}{x^2} dx,$$

which shows that the integral converges as $n \to \infty$.

So $\log(n!) = \left(n + \frac{1}{2}\right) - n + c_n$ where $c_n \to c$ for some value c as $n \to \infty$. So

$$n! \sim Ce^{n+\frac{1}{2}} e^{-n}$$

for some constant C.

We have shown that

$$n! \sim Ce^{n+\frac{1}{2}} e^{-n}$$

for some constant C.

Stirling's formula says that $C = \sqrt{2\pi}$. But we can conclude this from what we already know. For if we go back and do our computations with a "general C" we will find that the only way we could get a probability (i.e. that the appropriate integrals are 1) is if $C = \sqrt{2\pi}$.

5.6.2 Euler's integral and Stirling's formula.

We will use Euler's Gamma function:

$$\Gamma(s) = \int_0^\infty t^{s-1} e^{-t} dt$$

so that

$$\Gamma(n+1) = n!.$$

We are going to make a change of variable in

$$\Gamma(s+1) = \int_0^\infty t^s e^{-t} dt.$$

Setting $t = s\tau$ this becomes

$$s^{s+1} \int_0^\infty \tau^s e^{-\tau s} d\tau = s^{s+1} e^{-s} \int_0^\infty e^{-s(\tau - 1 - \log \tau)} d\tau.$$

So

$$\Gamma(s+1) = s^{s+1} e^{-s} \int_0^\infty e^{-sf(\tau)} d\tau$$

where

$$f(\tau) = \tau - 1 - \log \tau.$$

Here is the idea: The function f vanishes at $\tau = 1$ and achieves a minimum there, tending to ∞ as $\tau \to 0$ or as $\tau \to \infty$. So for large values of s we expect that the contribution to the integral will come from τ near 1. Near 1 we will approximate f by its quadratic term which will then lead to Gaussian integral.

Here is the idea: The function f vanishes at $\tau = 1$ and achieves a minimum there, tending to ∞ as $\tau \to 0$ or as $\tau \to \infty$. So for large values of s we expect that the contribution to the integral will come from τ near 1. Near 1 we will approximate f by its quadratic term which will then lead to Gaussian integral.

The following details are taken from Courant & Hilbert *Methods of Mathematical Physics* Vol 1 pp. 522-524.

Let $0 < \epsilon < \frac{1}{2}$. For $\frac{1}{2} < \tau < 1$ we have

$$\tau - 1 - \log \tau = \int_\tau^1 \left(\frac{1}{u} - 1\right) du \geq \int_\tau^1 (1-u)du$$

$$= \frac{1}{2}(\tau - 1)^2 \geq \frac{1}{8}(\tau - 1)^2.$$

On the interval $(0, 1-\epsilon)$ the integrand $e^{-sf(\tau)}$ is less than its maximum value which is $e^{-s\epsilon^2/8}$. So

$$\int_0^{1-\epsilon} e^{-sf(\tau)} d\tau \leq e^{-s\epsilon^2/8}.$$

Similarly, for $1 \leq \tau \leq 4$

$$\tau - 1 - \log \tau = \int_1^\tau \left(1 - \frac{1}{u}\right) du \geq 1\frac{1}{4}\int_1^\tau (u-1)du = \frac{1}{8}(\tau - 1)^2.$$

Again replacing the integrand by its largest value shows that

$$\int_{1+\epsilon}^4 e^{-sf(\tau)} d\tau \leq 3e^{-s\epsilon^2/8}.$$

For $\tau \geq 4$, $\tau - 1 - \log \tau \geq \frac{3}{4}\tau - \log \tau \geq \frac{1}{4}\tau$. Hence for $s > 4$

$$\int_4^\infty e^{-sf(\tau)} d\tau < \int_4^\infty e^{-s\tau/4} d\tau < e^{-s} < e^{-s\epsilon^2/8}.$$

So if we take $\epsilon = s^{-2/5}$ we will have

$$e^s s^{-s-1}\Gamma(s+1) = \int_{1-\epsilon}^{1+\epsilon} e^{-sf(\tau)} d\tau + O(e^{-s^{1/5}/8}).$$

So we are left with the study of the integral $\int_{1-\epsilon}^{1+\epsilon} e^{-sf(\tau)} d\tau$. Now $f(1) = f'(1) = 0$ and $f''(1) = 1$. So we can write

$$f(\tau) = \frac{1}{2}(\tau - 1)^2 + (\tau - 1)^3\psi(\tau) \quad \text{where} \quad |\psi(\tau)| \leq M \text{ on } \frac{1}{2} \leq \tau \leq \frac{3}{2}.$$

So on this interval we have

$$e^{-s(\tau-1)^2/2}e^{-Ms^{-1/5}} \leq e^{-sf(\tau)} \leq e^{-s(\tau-1)^2/2}e^{Ms^{-1/5}}$$

and

$$e^{-sf(\tau)} = e^{-s(\tau-1)^2/2}(1+O(s^{-1/5})).$$

So

$$\int_{1-\epsilon}^{1+\epsilon} e^{-sf(\tau)}d\tau = (1+O(s^{-1/5}))\int_{-\epsilon}^{\epsilon} e^{-su^2/2}du.$$

We have shown that

$$\int_{1-\epsilon}^{1+\epsilon} e^{-sf(\tau)}d\tau = (1+O(s^{-1/5}))\int_{-\epsilon}^{\epsilon} e^{-su^2/2}du.$$

If we make the change of variables $v = s^{\frac{1}{2}}u$ the integral on the right becomes

$$s^{-1/2}\int_{-\epsilon s^{\frac{1}{2}}}^{\epsilon s^{\frac{1}{2}}} e^{-v^2/2}dv$$

and

$$\int_{-\epsilon s^{\frac{1}{2}}}^{\epsilon s^{\frac{1}{2}}} e^{-v^2/2}dv \to \int_{-\infty}^{\infty} e^{-v^2/2}dv = \sqrt{2\pi}. \quad \square$$

Chapter 6

The contraction fixed point theorem.

Until now we have used the notion of metric quite informally. It is time for a formal definition. For any set X, we let $X \times X$ (called the Cartesian product of X with itself) denote the set of all ordered pairs of elements of X. (More generally, if X and Y are sets, we let $X \times Y$ denote the set of all pairs (x, y) with $x \in$ and $y \in Y$, and is called the Cartesian product of X with Y.)

6.1 Metrics and metric spaces.

A **metric** for a set X is a function d from $X \times X$ to the real numbers $\mathbb{R}$,

$$d : X \times X \to \mathbb{R}$$

such that for all $x, y, z \in X$

1. $d(x, y) = d(y, x)$
2. $d(x, z) \leq d(x, y) + d(y, z)$
3. $d(x, x) = 0$
4. If $d(x, y) = 0$ then $x = y$.

The inequality in 2) is known as the **triangle inequality** since if X is the plane and d the usual notion of distance, it says that the length of an edge of a triangle is at most the sum of the lengths of the two other edges. (In the plane, the inequality is strict unless the three points lie on a line.)

Condition 4. is in many ways inessential. It is often convenient to drop it, especially for the purposes of some proofs. For example, we might want to consider the decimal expansions $.49999\ldots$ and $.50000\ldots$ as different, but as having zero distance from one another. Or we might want to "identify" these two decimal expansions as representing the same point.

A function d which satisfies only conditions 1) - 3) is called a **pseudo-metric.**

A **metric space** is a pair (X, d) where X is a set and d is a metric on X. Almost always, when d is understood, we engage in the abuse of language and speak of "the metric space X".

Similarly for the notion of a **pseudo-metric space.**

In like fashion, we call $d(x, y)$ the **distance** between x and y, the function d being understood.

Open balls, the topology on a pseudo-metric space X.

If r is a positive number and $x \in X$, the (open) **ball of radius** r about x is defined to be the set of points at distance less than r from x and is denoted by $B_r(x)$. In symbols,

$$B_r(x) := \{y| \ d(x, y) < r\}.$$

If r and s are positive real numbers and if x and z are points of a pseudo-metric space X, it is possible that $B_r(x) \cap B_s(z) = \emptyset$. This will certainly be the case if $d(x, z) > r + s$ by virtue of the triangle inequality. Suppose that this intersection is not empty and that

$$w \in B_r(x) \cap B_s(z).$$

If $y \in X$ is such that $d(y, w) < \min[r - d(x, w), s - d(z, w)]$ then the triangle inequality implies that $y \in B_r(x) \cap B_s(z)$. Put another way, if we set $t := \min[r - d(x, w), s - d(z, w)]$ then

$$B_t(w) \subset B_r(x) \cap B_s(z).$$

Put still another way, this says that the intersection of two (open) balls is either empty or is a union of open balls. So if we call a set in X **open** if either it is empty, or is a union of open balls, we conclude that the intersection of any finite number of open sets is open, as is the union of any number of open sets. In technical language, we say that the open balls form a base for a topology on X.

Continuous maps and Lipschitz maps.

A map $f : X \to Y$ from one pseudo-metric space to another is called **continuous** if the inverse image under f of any open set in Y is an open set in X. Since an open set is a union of balls, this amounts to the condition that the inverse image of an open ball in Y is a union of open balls in X, or, to use the familiar ϵ,δ language, that if $f(x) = y$ then for every $\epsilon > 0$ there exists a $\delta = \delta(x, \epsilon) > 0$ such that

$$f(B_\delta(x)) \subset B_\epsilon(y).$$

Notice that in this definition δ is allowed to depend both on x and on ϵ. The map is called **uniformly continuous** if we can choose the δ independently of x.

An even stronger condition on a map from one pseudo-metric space to another is the **Lipschitz condition**. A map $f : X \to Y$ from a pseudo-metric space (X, d_X) to a pseudo-metric space (Y, d_Y) is called a **Lipschitz map** with **Lipschitz constant** C if

$$d_Y(f(x_1), f(x_2)) \leq C d_X(x_1, x_2) \quad \forall x_1, x_2 \in X.$$

Clearly a Lipschitz map is uniformly continuous.

An example. For example, suppose that A is a fixed subset of a pseudo-metric space X. Define the function $d(A, \cdot)$ from X to $\mathbb{R}$ by

$$d(A, x) := \inf\{d(x, w), \ w \in A\}.$$

The triangle inequality says that

$$d(x, w) \leq d(x, y) + d(y, w)$$

for all w, in particular for $w \in A$, and hence taking lower bounds we conclude that

$$d(A, x) \leq d(x, y) + d(A, y).$$

or

$$d(A, x) - d(A, y) \leq d(x, y).$$

Reversing the roles of x and y then gives

$$|d(A, x) - d(A, y)| \leq d(x, y).$$

Using the standard metric on the real numbers where the distance between a and b is $|a - b|$ this last inequality says that $d(A, \cdot)$ is a Lipschitz map from X to $\mathbb{R}$ with $C = 1$.

$$d(A, x) - d(A, y) \leq d(x, y).$$

Closed sets, and the closure of a set.

A **closed set** is defined to be a set whose complement is open. Since the inverse image of the complement of a set (under a map f) is the complement of the inverse image, we conclude that the inverse image of a closed set under a continuous map is again closed.

For example, the set consisting of a single point in $\mathbb{R}$ is closed. Since the map $d(A, \cdot)$ is continuous, we conclude that the set

$$\{x | d(A, x) = 0\}$$

consisting of all point at zero distance from A is a closed set. It clearly is a closed set which contains A. Suppose that S is some closed set containing A, and $y \notin S$.

Then there is some $r > 0$ such that $B_r(y)$ is contained in the complement of C, which implies that $d(y, w) \geq r$ for all $w \in S$. Thus $\{x|d(A, x) = 0\} \subset S$.

In short $\{x|d(A, x) = 0\}$ is a closed set containing A which is contained in all closed sets containing A. This is the definition of the **closure** of a set, which is denoted by $\overline{A}$. We have proved that

$$\overline{A} = \{x|d(A, x) = 0\}.$$

In particular, the closure of the one point set $\{x\}$ consists of all points u such that $d(u, x) = 0$.

Identifying points at zero distance.

The relation $d(x, y) = 0$ is an equivalence relation, call it R. (Transitivity being a consequence of the triangle inequality.) This then divides the space X into equivalence classes, where each equivalence class is of the form $\overline{\{x\}}$, the closure of a one point set. If $u \in \overline{\{x\}}$ and $v \in \overline{\{y\}}$ then

$$d(u, v) \leq d(u, x) + d(x, y) + d(y, v) = d(x, y).$$

since $x \in \overline{\{u\}}$ and $y \in \overline{\{v\}}$ we obtain the reverse inequality, and so

$$d(u, v) = d(x, y).$$

In other words, we may define the distance function on the quotient space X/R, i.e. on the space of equivalence classes by

$$d(\overline{\{x\}}, \overline{\{y\}}) := d(u, v), \quad u \in \overline{\{x\}}, v \in \overline{\{y\}}$$

and this does not depend on the choice of u and v. Axioms 1)-3) for a metric space continue to hold, but now

$$d(\overline{\{x\}}, \overline{\{y\}}) = 0 \Rightarrow \overline{\{x\}} = \overline{\{y\}}.$$

In other words, X/R is a *metric* space. Clearly the projection map $x \mapsto \overline{\{x\}}$ is an isometry of X onto X/R. (An isometry is a map which preserves distances.) In particular it is continuous. It is also open.

In short, we have provided a canonical way of passing (via an isometry) from a pseudo-metric space to a metric space by identifying points which are at zero distance from one another.

Dense subsets.

A subset A of a pseudo-metric space X is called *dense* if its closure is the whole space. From the above construction, the image A/R of A in the quotient space X/R is again dense. We will use this fact in the next section in the following form:

Proposition 6.1.1. *If $f : Y \to X$ is an isometry of Y such that $f(Y)$ is a dense set of X, then f descends to a map F of Y onto a dense set in the metric space X/R.*

6.2 Completeness and completion.

Cauchy sequences.

A sequence $\{y_n\}$ is said to be **Cauchy** if for any $\epsilon > 0$ there exists an $N = N(\epsilon$ such that

$$d(y_n, y_m) < \epsilon \quad \forall\ m, n > N.$$

The triangle inequality implies that every convergent sequence is Cauchy. But not every Cauchy sequence is convergent. For example, we can have a sequence of rational numbers which converge to an irrational number, as in the approximation to the square root of 2.

So if we look at the set of rational numbers as a metric space R in its own right, not every Cauchy sequence of rational numbers converges in R. We must "complete" the rational numbers to obtain $\mathbb{R}$, the set of real numbers. We want to discuss this phenomenon in general.

Complete metric spaces.

So we say that a (pseudo-)metric space is **complete** if every Cauchy sequence converges. The key result of this section is that we can always "complete" a metric or pseudo-metric space. More precisely, we claim that

Proposition 6.2.1. *Any metric (or pseudo-metric) space can be mapped by a one to one isometry onto a dense subset of a complete metric (or pseudo-metric) space.*

By Proposition 6.1.1, it is enough to prove this for a pseudo-metric space X.

Let X_{seq} denote the set of Cauchy sequences in X, and define the distance between the Cauchy sequences $\{x_n\}$ and $\{y_n\}$ to be

$$d(\{x_n\}, \{y_n\}) := \lim_{n\to\infty} d(x_n, y_n).$$

It is easy to check that d defines a pseudo-metric on X_{seq}. Let $f : X \to X_{seq}$ be the map sending x to the sequence all of whose elements are x;

$$f(x) = (x, x, x, x, \cdots).$$

It is clear that f is one to one and is an isometry. The image is dense since by definition

$$\lim d(f(x_n), \{x_n\}) = 0.$$

Since $f(X)$ is dense in X_{seq}, it suffices to show that any Cauchy sequence of points of the form $f(x_n)$ converges to a limit. But such a sequence converges to the element $\{x_n\}$. $\square$

6.2.1 Normed vector spaces.

Of special interest are vector spaces which have a metric which is compatible with the vector space properties and which is complete: Let V be a vector space over the real numbers. A **norm** is a real valued function

$$v \mapsto \|v\|$$

on V which satisfies

1. $\|v\| \geq 0$ and > 0 if $v \neq 0$,
2. $\|rv\| = |r|\|v\|$ for any real number r, and
3. $\|v + w\| \leq \|v\| + \|w\| \ \forall \ \ v, w \in V$.

Then $d(v, w) := \|v - w\|$ is a metric on V, which satisfies $d(v+u, w+u) = d(v, w)$ for all $v, w, u \in V$. The ball of radius r about the origin is then the set of all v such that $\|v\| < r$. A vector space equipped with a norm is called a **normed vector space** and if it is complete relative to the metric it is called a **Banach space**.

6.3 The contraction fixed point theorem.

We now come to the theorem which, despite the simplicity of its formulation and proof, should be considered as the central theorem of this book:

Let X and Y be metric spaces. Recall that a map $f : X \to Y$ is called a **Lipschitz map** or is said to be "Lipschitz continuous", if there is a constant C such that

$$d_Y(f(x_1), f(x_2)) \leq C d_X(x_1, x_2), \quad \forall \ x_1, x_2 \in X.$$

If f is a Lipschitz map, we may take the greatest lower bound of the set of all C for which the previous inequality holds. The inequality will continue to hold for this value of C which is known as the Lipschitz constant of f and denoted by $\mathrm{Lip}(f)$.

A map $K : X \to Y$ is called a **contraction** if it is Lipschitz, and its Lipschitz constant satisfies $\mathrm{Lip}(K) < 1$.

Suppose $K : X \to X$ is a contraction, and suppose that $Kx_1 = x_1$ and $Kx_2 = x_2$. Then

$$d(x_1, x_2) = d(Kx_1, Kx_2) \leq \mathrm{Lip}(K) d(x_1, x_2)$$

which is only possible if $d(x_1, x_2) = 0$, i.e. $x_1 = x_2$. So a contraction can have *at most one fixed point.* The contraction fixed point theorem asserts that if the metric space X is complete (and non-empty) then such a fixed point exists.

Theorem 6.3.1. [The contraction fixed point theorem.] *Let X be a non-empty complete metric space and $K : X \to X$ a contraction. Then K has a unique fixed point.*

Proof. Choose any point $x_0 \in X$ and define

$$x_n := K^n x_0$$

so that

$$x_{n+1} = Kx_n, \quad x_n = Kx_{n-1}$$

and therefore

$$d(x_{n+1}, x_n) \leq Cd(x_n, x_{n-1}), \quad 0 \leq C < 1$$

implying that

$$d(x_{n+1}, x_n) \leq C^n d(x_1, x_0).$$

Thus for any $m > n$ we have

$$d(x_m, x_n) \leq \sum_{n}^{m-1} d(x_{i+1}, x_i) \leq \left(C^n + C^{n+1} + \cdots + C^{m-1}\right) d(x_1, x_0)$$

$$\leq C^n \frac{d(x_1, x_0)}{1 - C}.$$

This says that the sequence $\{x_n\}$ is Cauchy. Since X is complete, it must converge to a limit x, and $Kx = \lim Kx_n = \lim x_{n+1} = x$ so x is a fixed point. We already know that this fixed point is unique. □

6.3.1 Local contractions.

We often encounter mappings which are contractions only near a particular point p. If K does not move p too much we can still conclude the existence of a fixed point, as in the following:

Theorem 6.3.2. *Let D be a closed ball of radius r centered at a point p in a complete metric space X, and suppose $K : D \to X$ is a contraction with Lipschitz constant $C < 1$. Suppose that*

$$d(p, Kp) \leq (1 - C)r.$$

Then K has a unique fixed point in D.

Proof. We simply check that $K : D \to D$ and then apply the preceding theorem with X replaced by D: For any $x \in D$, we have

$$d(Kx, p) \leq d(Kx, Kp) + d(Kp, p) \leq$$

$$Cd(x, p) + (1 - C)r \leq Cr + (1 - C)r = r.$$

□

Here is another version:

Theorem 6.3.3. *Let B be an open ball or radius r centered at p in a complete metric space X and let $K : B \to X$ be a contraction with Lipschitz constant $C < 1$. Suppose that*

$$d(p, Kp) < (1 - C)r.$$

Then K has a unique fixed point in B.

Proof. Restrict K to any slightly smaller closed ball centered at p and apply the preceding theorem. □

Estimating the distance to the fixed point.

Corollary 6.3.1. *Let $K : X \to X$ be a contraction with Lipschitz constant C of a complete metric space. Let x be its (unique) fixed point. Then for any $y \in X$ we have*

$$d(y, x) \leq \frac{d(y, Ky)}{1 - C}.$$

Proof. We may take $x_0 = y$ and follow the proof of Theorem 6.3.1. Alternatively, we may apply Prop. 6.3.2 to the closed ball of radius $d(y, Ky)/(1 - C)$ centered at y. Prop. 6.3.2 implies that the fixed point lies in the ball of radius r centered at y. □

Future applications of the Corollary.

The Corollary we just proved will be of use to us in proving continuous dependence on a parameter in the next section

Later, when we study iterative function systems for the construction of fractal images, the corollary becomes the "collage theorem". We might call our corollary the "abstract collage theorem".

6.4 Dependence on a parameter.

Suppose that the contraction "depends on a parameter s". More precisely, suppose that S is some other metric space and that

$$K : S \times X \to X$$

with

$$d_X(K(s, x_1), K(s, x_2)) \leq C d_X(x_1, x_2), \quad 0 \leq C < 1, \ \forall s \in S, \ x_1, x_2 \in X. \tag{6.1}$$

(We are assuming that the C in this inequality does not depend on s.) If we hold $s \in S$ fixed, we get a contraction

$$K_s : X \to X, \qquad K_s(x) := K(s, x).$$

This contraction has a unique fixed point, call it p_s. We thus obtain a map

$$S \to X, \qquad s \mapsto p_s$$

sending each $s \in S$ into the fixed point of K_s.

Theorem 6.4.1. [Continuous dependence of the fixed point on the parameter.] *Suppose that for each fixed $x \in X$, the map*

$$s \mapsto K(s, x)$$

of $S \to X$ is continuous. Then the map

$$s \mapsto p_s$$

is continuous.

Proof. Fix a $t \in S$ and an $\epsilon > 0$. We must find a $\delta > 0$ such that $d_X(p_s, p_t) < \epsilon$ if $d_S(s,t) < \delta$. Our continuity assumption says that we can find a $\delta > 0$ such that

$$d_X(K(s, p_t), p_t) = d_X(K(s, p_t), K(t, p_t) \leq (1 - C)\epsilon$$

if $d_S(s,t) < \delta$. This says that K_s moves p_t a distance at most $(1 - C)\epsilon$. But then the "abstract collage theorem", Prop. 6.3.1, says that

$$d_X(p_t, p_s) \leq \epsilon.$$

□

6.5 The Lipschitz implicit function theorem.

In this section we follow the treatment in [Shub]

6.5.1 The inverse function theorem.

We begin with the inverse function theorem whose proof contains the guts of the argument.

Consider a map $F : B_r(0) \to E$ where $B_r(0)$ is the open ball of radius r about the origin in a Banach space, E, and where $F(0) = 0$. Under suitable conditions on F, wish to conclude the existence of an inverse to F, defined on a possible smaller ball by means of the contraction fixed point theorem.

For example, suppose that F is continuously differentiable with derivative A at the origin which is invertible. Replacing F by $A^{-1}F$ we may assume that the derivative of F at 0 is the identity map, id.

So $F - \mathrm{id}$ vanishes at the origin together with its derivative. Hence the mean value theorem implies that we can arrange that $F - \mathrm{id}$ has Lipschitz constant as small as we like. This justifies the hypotheses of the following theorem:

Theorem 6.5.1. *Let $F : B_r(0) \to E$ satisfy $F(0) = 0$ and*

$$\mathrm{Lip}[F - \mathrm{id}] = \lambda < 1. \tag{6.2}$$

Then the ball $B_s(0)$ is contained in the image of F where

$$s = (1-\lambda)r \tag{6.3}$$

and F has an inverse, G defined on $B_s(0)$ with

$$\mathrm{Lip}[G - \mathrm{id}] \leq \frac{\lambda}{1-\lambda}. \tag{6.4}$$

Proof. Let us set $F = \mathrm{id} + v$ so

$$\mathrm{id} + v : B_r(0) \to E, \quad v(0) = 0, \quad \mathrm{Lip}[v] < \lambda < 1.$$

We want to find a $w : B_s(0) \to E$ with

$$w(0) = 0$$

and

$$(\mathrm{id} + v) \circ (\mathrm{id} + w) = \mathrm{id}.$$

This equation is the same as

$$w = -v \circ (\mathrm{id} + w).$$

Let X be the space of continuous maps of $\overline{B_s(0)} \to E$ satisfying

$$u(0) = 0$$

and

$$\mathrm{Lip}[u] \leq \frac{\lambda}{1-\lambda}.$$

Then X is a complete metric space relative to the sup norm, and, for $x \in \overline{B_s(0)}$ and $u \in X$ we have

$$\|u(x)\| = \|u(x) - u(0)\| \leq \frac{\lambda}{1-\lambda}\|x\| \leq r.$$

Thus, if $u \in X$ then

$$u : \overline{B_s} \to \overline{B_r}.$$

If $w_1, w_2 \in X$,

$$\| -v \circ (\mathrm{id} + w_1) + v \circ (\mathrm{id} + w_2) \| \leq \lambda \| (\mathrm{id} + w_1) - (\mathrm{id} + w_2) \| = \lambda \| w_1 - w_2 \| .$$

So the map $K : X \to X$

$$K(u) = -v \circ (\mathrm{id} + u)$$

is a contraction. Hence there is a unique fixed point. This proves the inverse function theorem. □

6.5.2 The implicit function theorem.

The setup.

We want to solve the equation $F(x,y) = 0$ for y as a function of x. In other words, we are looking for a function $y = g(x)$ such that $F(x, g(x)) \equiv 0$.

Here x and y are vector variables, say x ranges over an open ball A in a Banach space X and y ranges over and open ball B in some other Banach space Y.

Here $F : A \times B \to Z$ where Z is some third vector space.

To keep the notation simple, we will assume that A and B are open balls about the origin(s) and that

$$F(0,0) = 0.$$

The assumptions about F.

We assume that

- F is continuous as a function of (x,y).
- $\frac{\partial F}{\partial y}$ exists and is continuous as a function of (x,y). Remember that $\frac{\partial F}{\partial y}$ is a linear transformation. For example, it is a matrix if B is some ball in $\mathbb{R}^n$.
- $\frac{\partial F}{\partial y}(0,0)$ is invertible.

The method.

We set $T := \frac{\partial F}{\partial y}(0,0)$ and define

$$K(x,y) := y - T^{-1}F(x,y).$$

So K is a continuous map from $A \times B \to Y$ and

$$K(x,y)y = y \iff F(x,y) = 0.$$

K is a continuous map from $A \times B \to Y$ and $\frac{\partial K}{\partial y}$ is a continuous Y-valued function of (x,y) with

$$\frac{\partial K}{\partial y}(0,0) = 0.$$

So by choosing smaller balls (which we now rename as A and B) we can arrange that

$$\left\| \frac{\partial K}{\partial y}(x,y) \right\| < \frac{1}{2}$$

for all $(x,y) \in A \times B$. The mean value theorem implies that

K is a contraction in its second variable with Lipschitz constant $\frac{1}{2}$.

Let r denote the radius of the ball B. Since $K(x,y)-y$ is a continuous function of x and $K(0,0)-0=0$, we can shrink the ball A still further (and reuse A as the name of the shrunken ball) so that

$$\|K(x,y)-y\| < \frac{r}{2}$$

for all $(x,y) \in A \times B$.

We now recall a version of the contraction fixed point theorem with a slight change in notation suitable for our case:

Proposition 6.5.1. *Suppose that $K : A \times B \to Y$ is continuous, satisfies*

$$\|K(x,y_1) - K(x,y_2)\| \leq \|y_1 - y_2\|C, \qquad 0 \leq C < 1, \ \forall s \in S,$$

and

$$\|K(x,y) - y\| < (1-C)r, \quad \forall \ x \in A.$$

Then for each $x \in A$ there is a unique $y_x \in B$ such that $K(x,y) = y_x$, and the map $x \mapsto y_x$ is continuous.

We have verified the hypotheses of the proposition with $C = \frac{1}{2}$. So we have proved:

Theorem 6.5.2. [The Lipschitz implicit function theorem.] *Let $(x,y) \mapsto F(x,y) \in Z$ be a continuous map defined on an open set of $X \times Y$ where X, Y, and Z are Banach spaces and such that $\frac{\partial F}{\partial y}$ is continuous. Suppose that $F(x_0,y_0) = 0$ at some point (x_0,y_0) and $\frac{\partial F}{\partial y}(x_0,y_0)$ is invertible. Then there are open balls A and B about x_0 and y_0 such that for each $x \in A$ there is a unique $y = g(x) \in B$ such that $F(x,y) = 0$. The map g so defined is continuous.*

By the arguments we gave in Chapter I, we know that if F is continuously differentiable in both variables then g is differentiable near x_0 and we know how to compute its derivative.

6.6 The local existence theorem for solutions of differential equations.

The set up.

A is an open subset of a Banach space X, $I \subset \mathbb{R}$ is an open interval and

$$F : I \times A \to X$$

is continuous. We want to study the differential equation

$$\frac{dx}{dt} = F(t,x).$$

A **solution** of this equation is a map $f : J \to A$, where J is an open subinterval of I such that $f'(t)$ exists for all $t \in J$ and

$$f'(t) = F(t, f(t)).$$

If f' exists then f must be continuous, and the the right hand side of the above equation is then continuous. So any solution must be continuously differentiable.

The hypothesis.

The function F is uniformly Lipschitz in the second variable. That is, there is a constant c independent of t such that

$$\|F(t, x_1) - F(t, x_2)\| \leq c\|x_1 - x_2\| \quad \forall\, x_1, x_2 \in A.$$

The conclusion.

Theorem 6.6.1. *For any $(t_0, x_0) \in I \times A$ there is a neighborhood U of x_0 such that for any sufficiently small interval J containing t_0 there is a unique map $f : J \to U$ such that f is a solution to the differential equation and*

$$f(t_0) = x_0.$$

The idea of the proof.

If f is a solution to our differential equation defined on the interval J, then

$$f(t) - f(t_0) = \int_{t_0}^{t} F(s, f(s))ds.$$

If $f(t_0) = x_0$ then we get

$$f(t) = x_0 + \int_{t_0}^{t} F(s, f(s))ds.$$

Conversely, if f satisfies this last equation, then $f(t_0) = x_0$ and also f is differentiable and is a solution to our differential equation.

So for any interval J about t_0 let $\mathbb{B}(J)$ denote the space of **bounded, continuous** maps from J to X and try to define the map

$$K : \mathbb{B}(J) \to \mathbb{B}(J)$$

by

$$K(g)(t) = x_0 + \int_{t_0}^{t} F(s, g(s))ds.$$

If we can arrange by suitable choice of U that for small enough J the map K is defined and is a contraction, then the fixed point theorem gives us a unique solution to our differential equation with initial condition.

The proof.

Choose U to be a ball of radius r about x_0 and an interval L about t_0 so that F is bounded on $L \times U$ with bound m. Recall that c is the Lipschitz constant of F in its second variable.

Let $\mathbf{x_0}$ denote the constant function of t with value x_0 i.e.

$$\mathbf{x_0}(t) \equiv x_0.$$

Let B_r be the ball of radius r in $\mathbb{B}(J)$ about $\mathbf{x_0}$ relative to the sup norm

$$\|g - h\|_\infty := l.u.b._{t \in J}\|g(t) - h(t)\|.$$

So for any $g \in B_r$, $g(t) \in U$ for all $t \in J$ and so $F(t, g(t))$ is defined.

Let δ denote the length of J. If $g_1, g_2 \in B_r$ then $K(g_1)$ and $K(g_2)$ are defined, and for $t \in J$ we have

$$\|K(g_1)(t) - K(g_2)(t)\| = \left\| \int_{t_0}^{t} (F(s, g_1(s)) - F(s, g_2(s)))\, ds \right\|$$

$$\leq c\delta \|g_1 - g_2\|_\infty.$$

Taking the least upper bound with respect to t gives

$$\|Kg_1 - Kg_2\|_\infty \leq c\delta\|g_1 - g_2\|_\infty.$$

In other words, K is Lipschitz with Lipschitz constant $C = c\delta$. We need to choose δ so that $C = c\delta < 1$ if we want K to be a contraction.

Now let's see how far K moves the center, $\mathbf{x_0}$ of B_r: We have

$$\|K(\mathbf{x_0})(t) - \mathbf{x_0}(t)\| = \left\| \int_{t_0}^{t} F(s, x_0) ds \right\| \leq \delta m$$

so

$$\|K(\mathbf{x_0}) - \mathbf{x_0}\|_\infty \leq \delta m.$$

We will apply Theorem 6.3.2. So we want to choose δ small enough that

$$\delta m \leq (1 - C)r = (1 - \delta c)r.$$

The condition $\delta m \leq (1 - \delta c)r$ translates into $\delta(m + cr) \leq 1$ so if we choose

$$\delta < \frac{r}{m + cr}$$

then $C = \delta c < 1$ and $\|K(\mathbf{x_0}) - \mathbf{x_0}\|_\infty < (1 - C)r$ so K satisfies the conditions of the theorem and there is a unique fixed point. □

Chapter 7

The Hausdorff metric and Hutchinson's theorem.

7.1 The Hausdorff metric.

Compact sets and their collars.

Let X be a complete metric space. Let $\mathcal{H}(X)$ denote the space of non-empty compact subsets of X. For any $A \in \mathcal{H}(X)$ and any positive number ϵ, let

$$A_\epsilon = \{x \in X | d(x, y) \le \epsilon, \text{for some } y \in A\}.$$

We call A_ϵ the **ϵ-collar** of A. Recall that we defined

$$d(x, A) = \inf_{y \in A} d(x, y)$$

to be the distance from any $x \in X$ to A. So we can write the definition of the ϵ-collar as

$$A_\epsilon = \{x | d(x, A) \le \epsilon\}.$$

Notice that the infimum in the definition of $d(x, A)$ is actually achieved, that is, there is some point $y \in A$ such that

$$d(x, A) = d(x, y).$$

This is because A is compact. For a pair of non-empty compact sets, A and B, define

$$d(A, B) = \max_{x \in A} d(x, B).$$

So

$$d(A, B) \le \epsilon \iff A \subset B_\epsilon.$$

Notice that this condition is *not* symmetric in A and B.

The Hausdorffmetric $h(A, B)$**.**

So Hausdorff introduced

$$\begin{aligned} h(A,B) &= \max\{d(A,B), d(B,A)\} &(7.1)\\ &= \inf\{\epsilon \mid A \subset B_\epsilon \text{ and } B \subset A_\epsilon\}. &(7.2)\end{aligned}$$

as a distance on $\mathcal{H}(X)$. He proved

Theorem 7.1.1. *The function h on* $\mathcal{H}(X) \times \mathcal{H}(X)$ *satsifies the axioms for a metric and makes* $\mathcal{H}(X)$ *into a complete metric space. Furthermore, if*

$$A, B, C, D \in \mathcal{H}(X)$$

then

$$h(A \cup B, C \cup D) \leq \max\{h(A, C), h(B, D)\}. \tag{7.3}$$

Proof of (7.3). If ϵ is such that $A \subset C_\epsilon$ and $B \subset D_\epsilon$ then clearly $A \cup B \subset C_\epsilon \cup D_\epsilon = (C \cup D)_\epsilon$. Repeating this argument with the roles of A, C and B, D interchanged proves (7.3). □

We prove that h is a metric: h is symmetric, by definition. Also, $h(A, A) = 0$, and if $h(A, B) = 0$, then every point of A is within zero distance of B,and hence must belong to B since B is compact, so $A \subset B$ and similarly $B \subset A$. So $h(A, B) = 0$ implies that $A = B$.

Proof of the triangle inequality. For this it is enough to prove that

$$d(A, B) \leq d(A, C) + d(C, B),$$

because interchanging the role of A and B gives the desired result. Now for any $a \in A$ we have

$$\begin{aligned} d(a, B) &= \min_{b \in B} d(a, b)\\ &\leq \min_{b \in B} (d(a, c) + d(c, b)) \ \forall c \in C\\ &= d(a, c) + \min_{b \in B} d(c, b) \ \forall c \in C\\ &= d(a, c) + d(c, B) \ \forall c \in C\\ &\leq d(a, c) + d(C, B) \ \forall c \in C.\end{aligned}$$

The second term in the last expression does not depend on c, so minimizing over c gives

$$d(a, B) \leq d(a, C) + d(C, B).$$

Maximizing over a on the right gives

$$d(a, B) \leq d(A, C) + d(C, B).$$

Maximizing on the left gives the desired

$$d(A, B) \leq d(A, C) + d(C, A). \quad \square$$

bfA sketch of the proof of completeness. Let A_n be a sequence of compact non-empty subsets of X which is Cauchy in the Hausdorff metric. Define the set A to be the set of all $x \in X$ with the property that there exists a sequence of points $x_n \in A_n$ with $x_n \to x$. It is straighforward to prove that A is compact and non-empty and is the limit of the A_n in the Hausdorffmetric.

7.1.1 Contractions and the Hausdorffmetric.

Suppose that $K : X \to X$ is a contraction. Then K defines a transformation on the space of subsets of X (which we continue to denote by K):

$$K(A) = \{Kx | x \in A\}.$$

Since K continuous, it carries $\mathcal{H}(X)$ into itself. Let c be the Lipschitz constant of K. Then

$$\begin{aligned} d(K(A), K(B)) &= \max_{a \in A}[\min_{b \in B} d(K(a), K(b))] \\ &\leq \max_{a \in A}[\min_{b \in B} cd(a, b)] \\ &= cd(A, B). \end{aligned}$$

Similarly, $d(K(B), K(A)) \leq c\, d(B, A)$ and hence

$$h(K(A), K(B)) \leq c\, h(A, B). \tag{7.4}$$

In other words, *a contraction on X induces a contraction on $\mathcal{H}(X)$ with the same Lipschitz constant.*

7.2 Hutchinson's theorem.

The previous remark together with the following observation is the key to Hutchinson's remarkable construction of fractals:

Proposition 7.2.1. *Let $T_1, \ldots, T_n$ be a collection of contractions on $\mathcal{H}(X)$ with Lipschitz constants $c_1, \ldots, c_n$, and let $c = \max c_i$. Define the transformation T on $\mathcal{H}(X)$ by*

$$T(A) = T_1(A) \cup T_2(A) \cup \cdots \cup T_n(A).$$

Then T is a contraction with Lipschitz constant c.

Proof. By induction, it is enough to prove this for the case $n = 2$. By (7.3) we have

$$\begin{aligned} h(T(A), T(B)) &= h(T_1(A) \cup T_2(A), T_1(B) \cup T_2(B)) \\ &\leq \max\{h(T_1(A), T_1(B)), h(T_2(A), T_2(B))\} \\ &\leq \max\{c_1 h(A, B), c_2 h(A, B)\} \\ &= h(A, B) \max\{c_1, c_2\} = c \cdot h(A, B) \end{aligned}$$

. ∎

Putting the previous facts together we get **Hutchinson's theorem:**

Theorem 7.2.1. *Let $K_1, \ldots, K_n$ be contractions on a complete metric space and let c be the maximum of their Lifschitz contants. Define the Hutchinson operator, K, on $\mathcal{H}(X)$ by*

$$K(A) = K_1(A) \cup \cdots \cup K_n(a).$$

Then K is a contraction with Lipschtz constant c.

7.3 Affine examples.

In this section we describe several examples in which X is a subset of a vector space and each of the T_i in Hutchinson's theorem are affine transformations of the form

$$T_i : x \mapsto A_i x + b_i$$

where $b_i \in X$ and A_i is a linear transformation.

7.3.1 The classical Cantor set.

Take $X = [0, 1]$, the unit interval. Take

$$T_1 : x \mapsto \frac{x}{3}, \quad T_2 : x \mapsto \frac{x}{3} + \frac{2}{3}.$$

These are both contractions, so by Hutchinson's theorem there exists a unique closed set C invariant under T. This is the Cantor set.

Cantor's original construction.

To relate it to Cantor's original construction, let us go back to the proof of the contraction fixed point theorem applied to T acting on $\mathcal{H}(X)$. It says that if we start with any non-empty compact subset A_0 and keep applying T to it, i.e. set $A_n = T^n A_0$ then $A_n \to C$ in the Hausdorffmetric, h. Suppose we take the interval I itself as our A_0. Then

$$A_1 = T(I) = [0, \frac{1}{3}] \cup [\frac{2}{3}, 1].$$

in other words, applying the Hutchinson operator T to the interval $[0, 1]$ has the effect of deleting the "middle third" open interval $(\frac{1}{3}, \frac{2}{3})$. Applying T once more gives

$$A_2 = T^2[0, 1] = [0, \frac{1}{9}] \cup [\frac{2}{9}, \frac{1}{3}] \cup [\frac{2}{3}, \frac{7}{9}] \cup [\frac{8}{9}, 1].$$

In other words, A_2 is obtained from A_1 by deleting the middle thirds of each of the two intervals of A_1 and so on. This was Cantor's original construction. Since $A_{n+1} \subset A_n$ for this choice of initial set, the Hausdorff limit coincides with the intersection.

Using triadic expansions.

But of course Hutchinson's theorem (and the proof of the contractions fixed point theorem) says that we can start with *any* non-empty closed set as our initial "seed" and then keep applying T. For example, suppose we start with the one point set $B_0 = \{0\}$. Then $B_1 = TB_0$ is the two point set

$$B_1 = \{0, \frac{2}{3}\},$$

B_2 consists of the four point set

$$B_2 = \{0, \frac{2}{9}, \frac{2}{3}, \frac{8}{9}\}$$

and so on. We then must take the Hausdorff limit of this increasing collection of sets.

To describe the limiting set C from this point of view, it is useful to use triadic expansions of points in $[0, 1]$. Thus

$$\begin{aligned} 0 &= .0000000\cdots \\ 2/3 &= .2000000\cdots \\ 2/9 &= .0200000\cdots \\ 8/9 &= .2200000\cdots \end{aligned}$$

and so on. Thus the set B_n will consist of points whose triadic exapnsion has only zeros or twos in the first n positions followed by a string of all zeros.

Thus a point will lie in C (be the limit of such points) if and only if it has a triadic expansion consisting entirely of zeros or twos. This includes the possibility of an infinite string of all twos at the tail of the expansion. for example, the point 1 which belongs to the Cantor set has a triadic expansion $1 = .222222\cdots$. Similarly the point $\frac{2}{3}$ has the triadic expansion $\frac{2}{3} = .0222222\cdots$ and so is in the limit of the sets B_n. But a point such as $.101\cdots$ is not in the limit of the B_n and hence not in C. This description of C is also due to Cantor.

Since C (according to Cantor's second description) is closed, the uniqueness part of the fixed point theorem guarantees that the second description coincides with the first.

Self-similarity of the Cantor set.

Notice that for any point a with triadic expansion $a = .a_1a_2a_2\cdots$

$$T_1a = .0a_1a_2a_3\cdots, \quad \text{while} \quad T_2a = .2a_1a_2a_3\cdots.$$

Thus if all the entries in the expansion of a are either zero or two, this will also be true for T_1a and T_2a. This shows that the C (given by this second Cantor description) satisfies $TC \subset C$. On the other hand,

$$T_1(.a_2a_3\cdots) = .0a_2a_3\cdots, \quad T_2(.a_2a_3\cdots) = .2a_2a_3\cdots$$

which shows that $.a_1a_2a_3\cdots$ is in the image of T_1 if $a_1 = 0$ or in the image of T_2 if $a_1 = 2$. This shows that $TC = C$.

The statement that $TC = C$ is frequently formulated as saying that C is is **self-similar**, a notion emphasized and popularized by Mandelbrot.

More on this later.

7.3.2 The Sierpinski gasket.

Consider the three affine transformations of the plane:

$$T_1 : \begin{pmatrix} x \\ y \end{pmatrix} \mapsto \frac{1}{2}\begin{pmatrix} x \\ y \end{pmatrix}, \quad T_2 : \begin{pmatrix} x \\ y \end{pmatrix} \mapsto \frac{1}{2}\begin{pmatrix} x \\ y \end{pmatrix} + \frac{1}{2}\begin{pmatrix} 1 \\ 0 \end{pmatrix},$$

$$T_3 : \begin{pmatrix} x \\ y \end{pmatrix} \mapsto \frac{1}{2}\begin{pmatrix} x \\ y \end{pmatrix} + \frac{1}{2}\begin{pmatrix} 0 \\ 1 \end{pmatrix}.$$

The fixed point of the Hutchinson operator for this choice of T_1, T_2, T_3 is called the **Sierpinski gasket**, S.

If we take our initial set A_0 to be the right triangle with vertices at

$$\begin{pmatrix} 0 \\ 0 \end{pmatrix}, \begin{pmatrix} 1 \\ 0 \end{pmatrix}, \text{and} \quad \begin{pmatrix} 0 \\ 1 \end{pmatrix}$$

then each of the T_iA_0 is a similar right triangle whose linear dimensions are one-half as large, and which shares one common vertex with the original triangle.

In other words,

$$A_1 = TA_0$$

is obtained from our original triangle be deleting the interior of the (reversed) right triangle whose vertices are the midpoints of our original triangle. Just as in the case of the Cantor set, successive applications of T to this choice of original set amounts to sussive deletions of the "middle" and the Hausdorfflimit is the intersection of all them: $S = \bigcap A_i$.

We can also start with the one element set

$$B_0 \left\{ \begin{pmatrix} 0 \\ 0 \end{pmatrix} \right\}$$

Using a binary expansion for the x and y coordinates, application of T to B_0 gives the three element set

$$\left\{ \begin{pmatrix} 0 \\ 0 \end{pmatrix}, \begin{pmatrix} .1 \\ 0 \end{pmatrix}, \begin{pmatrix} 0 \\ .1 \end{pmatrix} \right\}.$$

The set $B_2 = TB_1$ will contain nine points, whose binary expansion is obtained from the above three by shifting the x and y expansions one unit to the right and either inserting a 0 before both expansions (the effect of T_1), insert a 1 before the expansion of x and a zero before the y or vice versa.

Proceeding in this fashion, we see that B_n consists of 3^n points which have all 0 in the binary expansion of the x and y coordinates, past the n-th position, and which are further constrained by the condition that at no earler point do we have both $x_i = 1$ and $y_1 = 1$. Passing to the limit shows that S consists of all points for which we can find (possible inifinite) binary expansions of the x and y coordinates so that $x_i = 1 = y_i$ never occurs.

For example $x = \frac{1}{2}, y = \frac{1}{2}$ belongs to S because we can write $x = .10000\cdots, y = .011111\ldots$. Again, from this (second) description of S in terms of binary expansions it is clear that $TS = S$.

7.3.3 A one line code for creating the Sierpinski gasket.

The following slide gives a matlab m file for doing the first seven approximations to the Sierpinksi gasket (in a slightly different orientation) as a movie. Notice that that iterative scheme is encoded in the single line

J=[J J;J zeros($2^i, 2^i$)];.

The other instructions are for the graphics, etc. This shows the power of Hutchinsons theorem.

```
J=[10];
image(J);colormap(colorcube(17))
pause(3)
for i=0:6
J=[J J;J zeros(2^i,2^i)];
image(J);
colormap(colorcube(17));
pause(3)
end
```

Here are the (first seven) successive images:

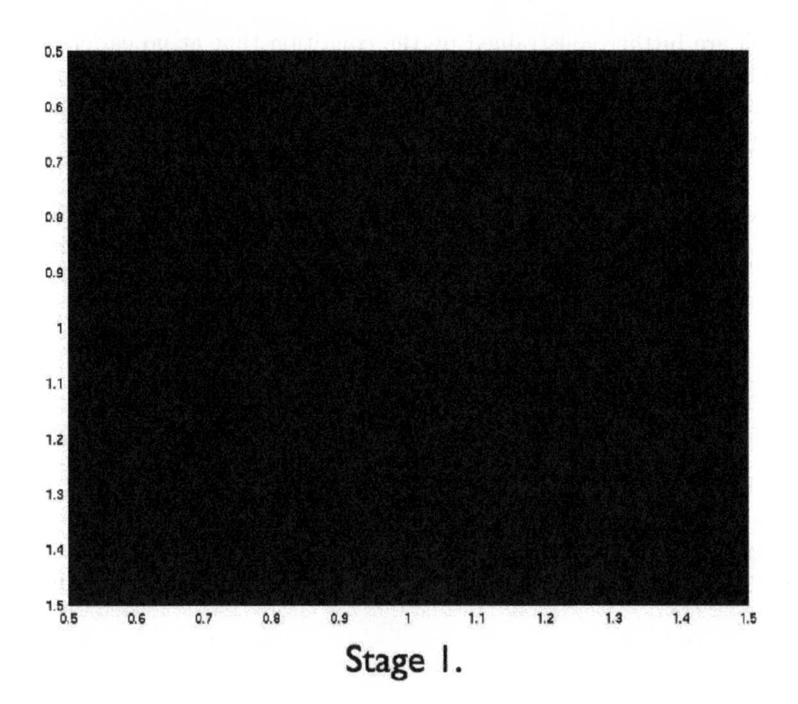

Stage 1.

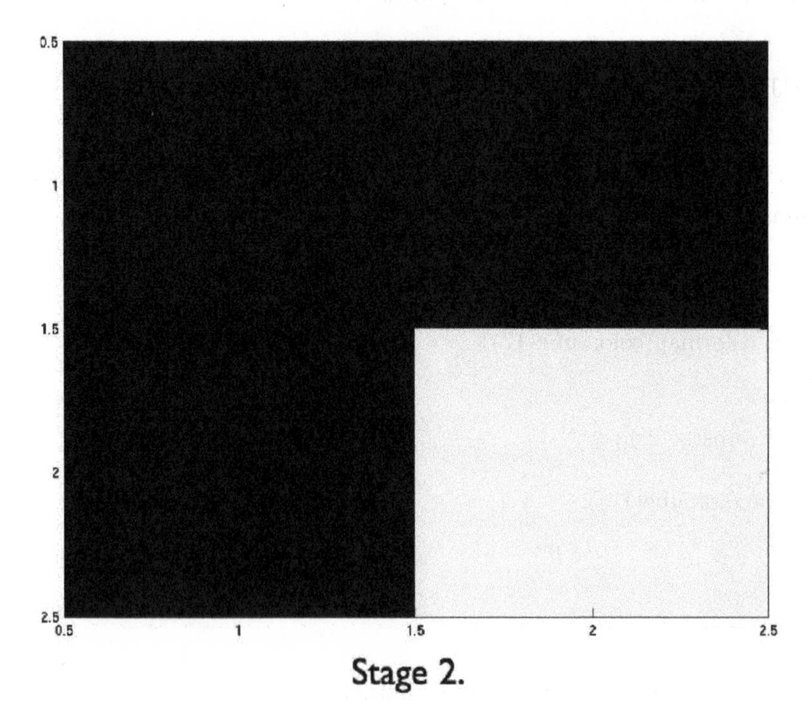

Stage 2.

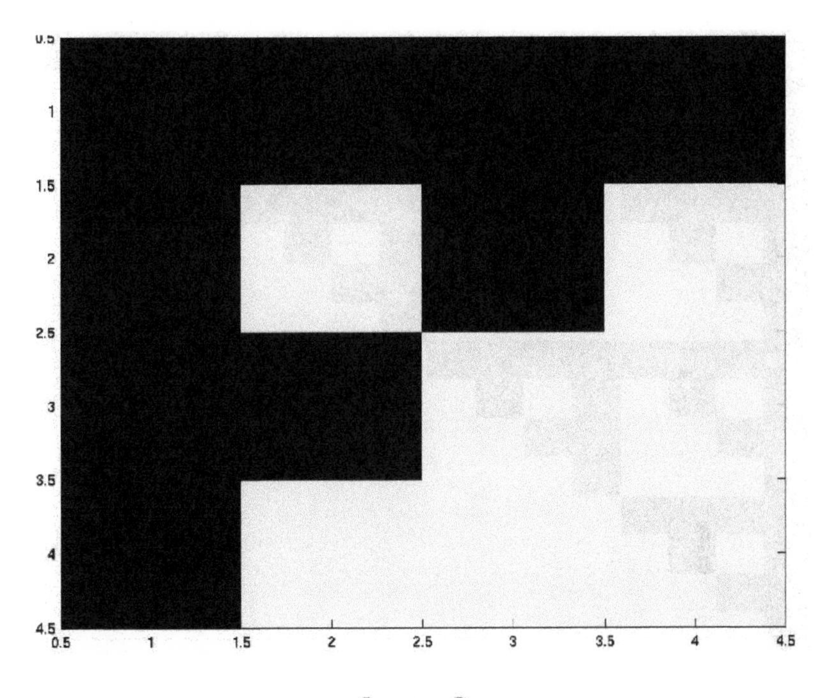

Stage 3.

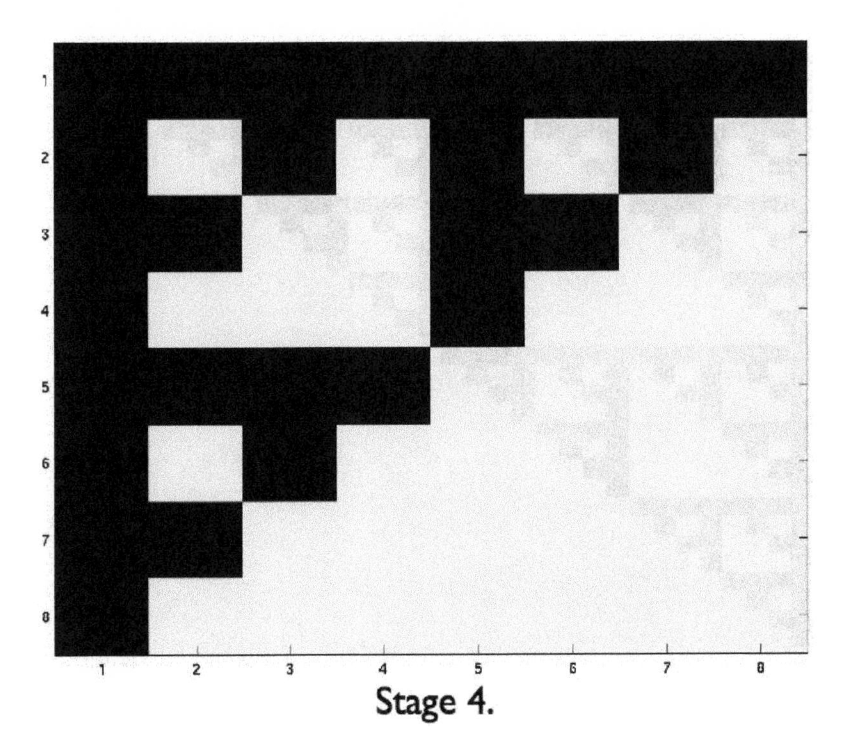

Stage 4.

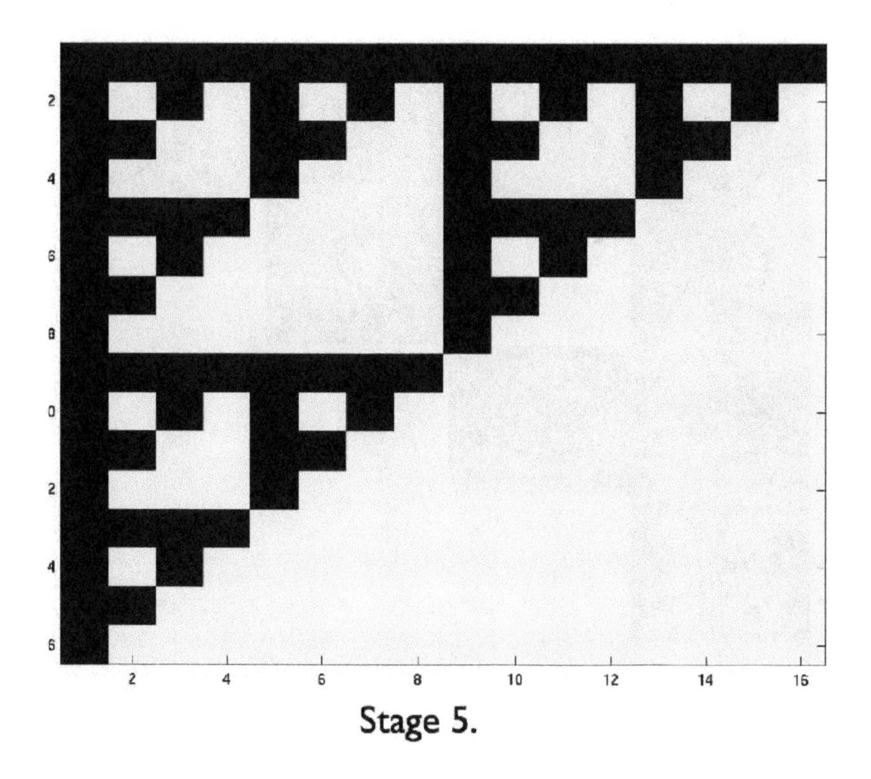

Stage 5.

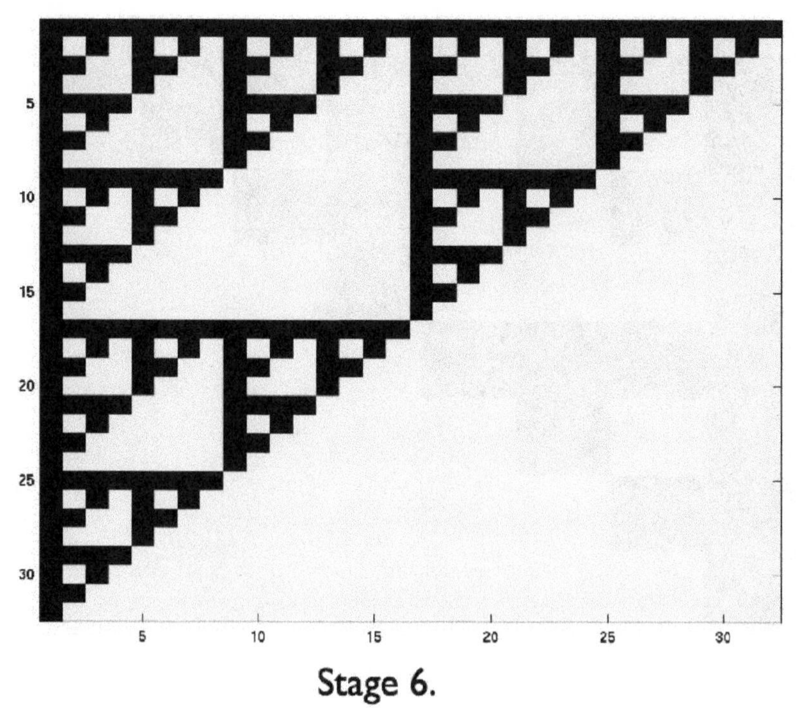

Stage 6.

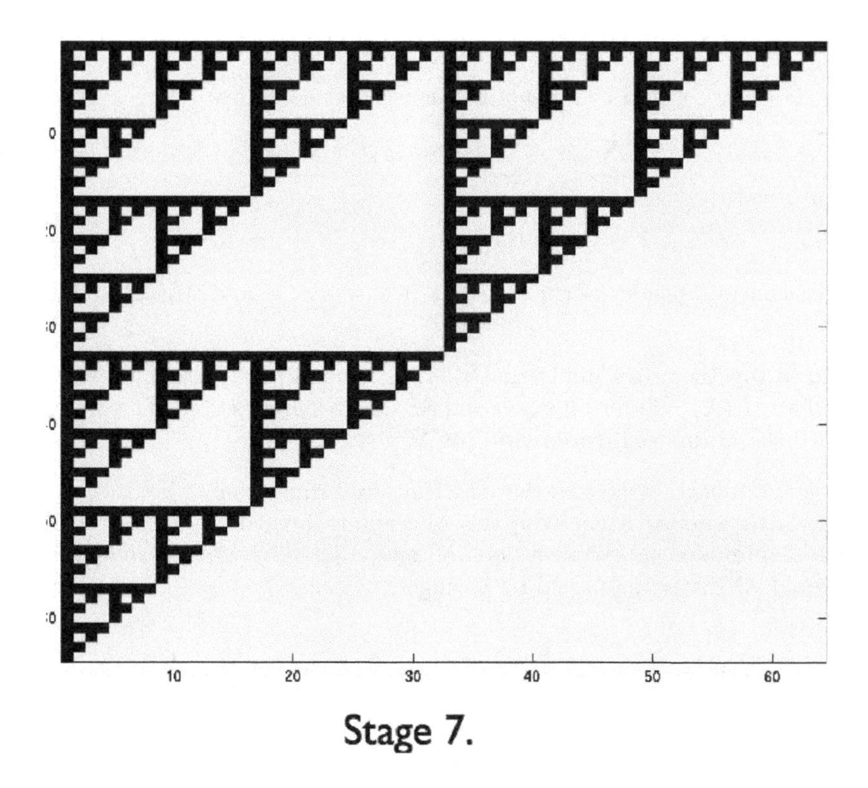

Stage 7.

7.4 Hausdorffdimension.

Let X be a metric space. Let $\mathcal{B}$ be a countable collection of balls B_i of radii $r_i < \epsilon$ which cover X, meaning that

$$X \subset \bigcup_i B_i.$$

(We assume that at least one such countable cover exists.)

For each non-negative real number t let

$$m_t(\mathcal{B}) := \sum_i r_i^t$$

where this sum might be infinite. Define

$$M_{t,\epsilon}(X) = g.l.b.\{m_t(\mathcal{B})\}$$

where we are taking the greatest lower bound over all possible countable covers of X by balls of radius at most ϵ. So $M_{t,\epsilon}(X)$ can be infinite, or be any non-negative real number, including 0. By its definition, $M_{t,\epsilon}$ is non-decreasing as $\epsilon \to 0$, so there is a limit (again possibly infinite or zero or any non-negative real number). Call this limit $M_t(X)$.

Theorem 7.4.1. *If* $0 < s < t$ *and* $M_s(X) < \infty$ *then* $M_t(X) = 0$.

Proof. For any cover $\mathcal{B}$ by balls of radius at most ϵ we have

$$\sum_i r_i^t = \sum_i r_i^{t-s} r_i^s \le \epsilon^{t-s} \sum_i r_i^s.$$

So $M_{t,\epsilon} \le \epsilon^{t-s} M_{s,\epsilon}$. Passing to the limit gives the theorem. □

The contrapositive assertion is that if $0 < s < t$ and $M_t(X) > 0$, then $M_s(X) = \infty$.

Taken together they imply that there is a number d (possibly zero or ∞) such that $M_t(X) = 0$ for all $t > d$ and $M_s(X) = \infty$ for $s < d$. This number d is called the **Hausdorffdimension** of X.

Easy arguments will show that the Hausdorff dimension of $\mathbb{R}^n$ is n.

For further information about this, and about the material in the remainder of this chapter, see the excellent book *Measure, Topology, and Fractal Geometry* by Gerald A. Edgar, published by Springer.

7.5 Similarity dimension of contracting ratio lists.

7.5.1 Contracting ratio lists.

A finite collection of real numbers

$$(r_1, \ldots, r_n)$$

is called a **contracting ratio list** if

$$0 < r_i < 1 \quad \forall\, i = 1, \ldots, n.$$

Theorem 7.5.1. *Let* $(r_1, \ldots, r_n)$ *be a contracting ratio list. There exists a unique non-negative real number* s *such that*

$$\sum_{i=1}^{n} r_i^s = 1. \tag{7.5}$$

The number s *is* 0 *if and only if* $n = 1$.

Proof. If $n = 1$ then $s = 0$ works and is clearly the only solution. If $n > 1$, define the function f on $[0, \infty)$ by

$$f(t) := \sum_{i=1}^{n} r_i^t.$$

We have $f(0) = n$ and $\lim_{t\to\infty} f(t) = 0 < 1$. Since f is continuous, there is some postive solution to (7.5). To show that this solution is unique, it is enough to show that f is monotone decreasing. This follows from the fact that its derivative is

$$\sum_{i=1}^{n} r_i^t \log r_i < 0.$$

□

Similarity dimension.

The number s in (7.5) is called the **similarity dimension** of the ratio list $(r_1, \ldots, r_n)$.

7.6 Iterated function systems and fractals.

A map $f : X \to Y$ between two metric spaces is called a **similarity** with similarity ratio r if

$$d_Y(f(x_1), f(x_2)) = r d_X(x_1, x_2) \;\; \forall\, x_1,\; x_2 \in X.$$

(Recall that a map is called **Lipschitz** with Lipschitz constant r if we only had an inequality, $\leq$, instead of equality in the above.)

7.6.1 Realizations of a contracting ratio list.

Let X be a complete metric space, and let $(r_1, \ldots, r_n)$ be a contracting ratio list. A collection

$$(f_1, \ldots, f_n), \quad f_i : X \to X$$

is called an **iterated function system** which **realizes** the contracting ratio list if

$$f_i : X \to X, \quad i = 1, \ldots, n$$

is a similarity with ratio r_i. We also say that $(f_1, \ldots, f_n)$ is a **realization** of the ratio list $(r_1, \ldots, r_n)$.

Using Hutchinson.

It is a consequence of *Hutchinson's theorem*, that

Theorem 7.6.1. *If $(f_1, \ldots, f_n)$ is a realization of the contracting ratio list $(r_1, \ldots, r_n)$ on a complete metric space, X, then there exists a unique non-empty compact subset $K \subset X$ such that*

$$K = f_1(K) \cup \cdots \cup f_n(K).$$

In fact, as we have seen, Hutchinson's theorem and the contraction fixed point theorem asserts the corresponding result where the f_i are merely assumed to be Lipschitz maps with Lipschitz constants $(r_1, \ldots, r_n)$ all < 1.

7.7 Fractals and fractal dimension.

The set K is sometimes called the **fractal** associated with the realization $(f_1, \ldots, f_n)$ of the contracting ratio list $(r_1, \ldots, r_n)$.

Here are some facts. For more details and the proofs, I refer to the book [?] mentioned above.

$$\dim(K) \leq s \tag{7.6}$$

where dim denotes Hausdorff dimension, and s is the similarity dimension of $(r_1, \ldots, r_n)$. In general, we can only assert an inequality here, for the the set K does not fix $(r_1, \ldots, r_n)$ or its realization.

For example, we can repeat some of the r_i and the corresponding f_i. This will give us a longer list, and hence a larger s, but will not change K. But we can demand a rather strong form of non-redundancy known as **Moran's condition**: There exists an open set O such that

$$O \supset f_i(O)\ \forall\ i \ \text{ and } \ f_i(O) \cap f_j(O) = \emptyset \ \ \forall\ i \neq j. \tag{7.7}$$

Then

Theorem 7.7.1. *If $(f_1, \ldots, f_n)$ is a realization of $(r_1, \ldots, r_n)$ on $\mathbb{R}^d$ and if Moran's condition holds then*

$$\dim K = s.$$

The Hausdorff dimension of the Cantor set and of the Sierpinski gasket.

In both of these examples the Moran condition is satisfied. So

- For the Cantor set we are looking for an s such that $\left(\frac{1}{3}\right)^s + \left(\frac{1}{3}\right)^s = 1$ which says that $\frac{2}{3^s} = 1$. Taking logarithms gives $\log 2 - s \log 3 = 0$ so the Hausdorff dimension of the Cantor set is $\log 2 / \log 3$.

- For the Sierpinski gasket the equation becomes $3 \cdot \frac{1}{2^s} = 1$ so the Hausdorff dimension of the Sierpinski gasket is $\log 3 / \log 2$.

Felix Hausdorff

Born: 8 Nov 1868 in Breslau, Germany (now Wroclaw, Poland)
Died: 26 Jan 1942 in Bonn, Germany by suicide, to avoid being sent to an extermination camp.

Chapter 8

Hyperbolicity.

Let E be a Banach space. A linear map $A : E \to E$ is called **hyperbolic** if we can find closed subspaces S and U of E which are invariant under A such that we have the direct sum decomposition

$$E = S \oplus U \tag{8.1}$$

and a positive constant $a < 1$ so that the estimates

$$\|A_s\| \leq a < 1, \quad A_s = A_{|S} \tag{8.2}$$

and

$$\|A_u^{-1}\| \leq a < 1, \quad A_u = A_{|U} \tag{8.3}$$

hold. (Here, as part of hypothesis (8.3), it is assumed that the restriction of A to U is an isomorphism so that A_u^{-1} is defined.)

If p is a fixed point of a diffeomorphism f, then it is called a **hyperbolic** fixed point if the linear transformation df_p is hyperbolic.

The main purpose of the first part of this chapter is to prove that any diffeomorphism, f is conjugate via a local *homeomorphism* to its derivative, df_p near a hyperbolic fixed point. A more detailed statement will be given below. We discussed the one dimensional version of this in Section 4.5. The theorem is a special case (C^0) of my old theorem about C^k conjugacy near a hyperbolic fixed point. See my paper "The structure of local homeomorphisms, II" *American Journal of Mathematics* vol **80** (1958) pp. 623-632. I will state (without proof) the C^∞ version later in this chapter.

8.1 The conjugacy theorem.

Our treatment will follow the exposition in [Shub].

8.1.1 A global version.

We begin with a global conjugacy theorem. We will then pass from the global to the local.

Theorem 8.1.1. *Let A be a hyperbolic isomorphism (so that A^{-1} is bounded) with a as above, and let*

$$\epsilon < \frac{1-a}{\|A^{-1}\|}. \tag{8.4}$$

If ϕ and ψ are bounded Lipschitz maps of E into itself with

$$\mathrm{Lip}[\phi] < \epsilon, \qquad \mathrm{Lip}[\psi] < \epsilon$$

then there is a unique solution to the equation

$$(\mathrm{id} + u) \circ (A + \phi) = (A + \psi) \circ (\mathrm{id} + u) \tag{8.5}$$

in the space, X of bounded continuous maps of E into itself. If $\phi(0) = \psi(0) = 0$ then $u(0) = 0$.

Proof. If we expand out both sides of (8.5) we get the equation

$$Au - u(A + \phi) = \phi - \psi(\mathrm{id} + u).$$

Let us define the linear operator, L, on the space X by

$$L(u) := Au - u \circ (A + \phi).$$

So we wish to solve the equation

$$L(u) = \phi - \psi(\mathrm{id} + u).$$

So we wish to solve the equation

$$L(u) := Au - u \circ (A + \phi). \tag{8.6}$$

We shall show that L is invertible with

$$\|L^{-1}\| \le \frac{\|A^{-1}\|}{(1-a)}. \tag{8.7}$$

Assume, for the moment that we have proved (8.7). We are then looking for a solution of

$$u = K(u)$$

where

$$K(u) = L^{-1}[\phi - \psi(\mathrm{id} + u)].$$

We have

$$\begin{aligned} \|K(u_1) - K(u_2)\| &= \|L^{-1}[\phi - \psi(\mathrm{id} + u_1) - \phi + \psi(\mathrm{id} + u_2)]\| \\ &= \|L^{-1}[\psi(\mathrm{id} + u_2) - \psi(\mathrm{id} + u_1)]\| \\ &\le \|L^{-1}\| \cdot \mathrm{Lip}[\psi] \cdot \|u_2 - u_1\| \\ &< c\|u_2 - u_1\|, \quad c < 1 \end{aligned}$$

if we combine (8.7) with (8.4). Thus K is a contraction and we may apply the contraction fixed point theorem to conclude the existence and uniqueness of the solution to (8.5).

So we must turn our attention to the proof that L is invertible and of the estimate (8.7). Let us write

$$Lu = A(Mu)$$

where

$$Mu = u - A^{-1}u \circ (A + \phi).$$

Composition with A is an invertible operator and the norm of its inverse is $\|A^{-1}\|$. So we are reduced to proving that M is invertible and that we have the estimate

$$\|M^{-1}\| \leq \frac{1}{1-a}. \tag{8.8}$$

Proof of 8.8. Let us write

$$u = f \oplus g, \quad f : E \to S, \quad g : E \to U$$

in accordance with the decomposition (8.1). If we let Y denote the space of bounded continuous maps from E to S, and let Z denote the space of bounded continuous maps from E to U, we have

$$X = Y \oplus Z$$

and the operator M sends each of the spaces Y and Z into themselves since A^{-1} preserves S and U.

Let M_s denote the restriction of M to Y, and let M_u denote the restriction of M to Z. It will be enough for us to prove that each of the operators M_s and M_u is invertible with a bound (8.8) with M replaced by M_s and by M_u. For $f \in Y$ let us write

$$M_s f = f - Nf, \quad Nf = A^{-1} f \circ (A + \phi).$$

We will prove:

Lemma 8.1.1. *The map N is invertible and we have*

$$\|N^{-1}\| \leq a.$$

We will break the proof of the lemma into several pieces:

- **The map $A + \phi$ is injective.** Indeed

$$\|Ax\| \geq \frac{1}{\|A^{-1}\|}\|x\|$$

so

$$\begin{aligned} \|Ax + \phi(x) - Ay - \phi(y)\| &\geq \left[\frac{1}{\|A^{-1}\|} - \mathrm{Lip}[\phi]\right]\|x - y\| \\ &\geq \frac{a}{\|A^{-1}\|}\|x - y\| \end{aligned}$$

by (8.4.

- **The map $A+\phi$ is surjective with continuous inverse.** To solve

$$Ax + \phi(x) = y$$

for x, we apply the contraction fixed point theorem to the map

$$x \mapsto A^{-1}(y - \phi(x)).$$

The estimate

$$\epsilon < \frac{1-a}{\|A^{-1}\|}. \qquad (8.4)$$

shows that this map is a contraction, so we can solve for x as function of y. Hence $A+\phi$ is surjective. In fact, this argument shows that *the map $A+\phi$ a homeomorphism with Lipschitz inverse.*

- Recall that the map N was defined by $Nf = A^{-1}f \circ (A+\phi)$.

  So the map N is invertible, with

$$N^{-1}f = A_s f \circ (A+\phi)^{-1}.$$

Since $\|A_s\| \leq a$, we have

$$\|N^{-1}f\| \leq a\|f\|.$$

(This is in terms of the sup norm on Y.) In other words, in terms of operator norms,

$$\|N^{-1}\| \leq a,$$

completing the proof of the lemma.

Recall that $\quad M_s f := f - Nf.$

We can now find M_s^{-1} by the geometric series

$$\begin{aligned} M_s^{-1} &= (I-N)^{-1} \\ &= [(-N)(I-N^{-1})]^{-1} \\ &= (-N)^{-1}[I + N^{-1} + N^{-2} + N^{-3} + \cdots] \end{aligned}$$

and so on Y we have the estimate

$$\|M_s^{-1}\| \leq \frac{a}{1-a}.$$

The restriction, M_u, of M to Z is

$$M_u g = g - Qg$$

with

$$\|Qg\| \leq a\|g\|$$

so we have the simpler series

$$M_u^{-1} = I + Q + Q^2 + \cdots$$

giving the estimate

$$\|M_u\| \le \frac{1}{1-a}.$$

Since

$$\frac{a}{1-a} < \frac{1}{1-a}$$

the two pieces together give the desired estimate

$$\|M\| \le \frac{1}{1-a},$$

completing the proof of the first part of the theorem.

Since evaluation at zero is a continuous function on X, to prove the last statement of the proposition it is enough to observe that if we start with an initial approximation satisfying $u(0) = 0$ (for example $u \equiv 0$) Ku will also satisfy this condition and hence so will $K^n u$ and therefore so will the unique fixed point.

This completes the proof of the theorem.

□

We now pass from the global to the local.

8.1.2 The local version.

Let f be a differentiable, diffeomorphism defined in some neighborhood of 0 with $f(0) = 0$ and $df_0 = A$ where A is hyperbolic. So we assume that A satisfies conditions (8.2) and (8.3) relative to a decomposition (8.1). We may write

$$f = A + \phi$$

where

$$\phi(0) = 0, \quad d\phi_0 = 0.$$

We wish to prove

Theorem 8.1.2. *There exists neighborhoods U and V of 0 and a homeomorphism $h : U \to V$ such that*

$$h \circ A = f \circ h. \tag{8.9}$$

In short, the theorem asserts that a diffeomorphism near a hyperbolic fixed point is conjugate to its linear part via a *homeomorphism.*

Plan of the proof.

We prove this theorem by modifying ϕ outside a sufficiently small neighborhood of 0 in such a way that the new ϕ is globally defined and has Lipschitz constant less than ϵ where ϵ satisfies condition (8.4). We can then apply the preceding theorem to find a global h which conjugates the modified f to A, and $h(0) = 0$. But since we will not have modified f near the origin, this will prove the local assertion of the theorem. For this purpose, choose some function $\rho : \mathbb{R} \to \mathbb{R}_+$ with

$$\begin{aligned} \rho(t) = 0 &\quad \forall \ t \geq 1 \\ \rho(t) = 1 &\quad \forall \ t \leq \frac{1}{2} \\ |\rho'(t)| &< K \ \forall \ t \end{aligned}$$

where K is some number,

$$K > 2.$$

For a fixed ϵ let r be sufficiently small so that the on the ball, $B_r(0)$ we have the estimate

$$\|d\phi_x\| < \frac{\epsilon}{2K},$$

which is possible since $d\phi_0 = 0$ and $d\phi$ is continuous.

The modification.

Define

$$\psi(x) := \rho\left(\frac{\|x\|}{r}\right)\phi(x)$$

for $\|x\| \leq r$, and continuously extend to all of E by setting

$$\psi(x) = 0, \quad \|x\| \geq r.$$

Notice that

$$\psi(x) = \phi(x), \quad \|x\| \leq \frac{r}{2}.$$

Checking the Lipschitz constant.

We check the Lipschitz constant of ψ. There are 3 alternatives: If x_1 and x_2 both belong to $B_r(0)$ we have $\|\psi(x_1) - \psi(x_2)\|$

$$\begin{aligned} &= \left\|\rho\left(\frac{\|x_1\|}{r}\right)\phi(x_1) - \rho\left(\frac{\|x_2\|}{r}\right)\phi(x_2)\right\| \\ &\leq \left|\rho\left(\frac{\|x_1\|}{r}\right) - \rho\left(\frac{\|x_2\|}{r}\right)\right| \|\phi(x_1)\| + \rho\left(\frac{\|x_2\|}{r}\right)\|\phi(x_1) - \phi(x_2)\| \\ &\leq (K\|x_1 - x_2\|/r) \times \|x_1\| \times (\epsilon/2K) + (\epsilon/2K) \times \|x_1 - x_2\| \\ &\leq \epsilon\|x_1 - x_2|. \end{aligned}$$

If $x_1 \in B_r(0)$, $x_2 \notin B_r(0)$, then the second term in the expression on the second line above vanishes and the first term is at most $(\epsilon/2)\|x_1 - x_2\|$. If neither x_1 nor x_2 belong to $B_r(0)$ then $\psi(x_1) - \psi(x_2) = 0 - 0 = 0$. We have verified that $\mathrm{Lip}[\psi] < \epsilon$ and so have proved the theorem. □

8.1.3 C^∞ conjugacy.

We started out with a diffeomorphism but only ended up with a conjugacy via a homeomorphism. Suppose we start out with an infinity differentiable diffeomorphism in Theorem 8.1.2. Can we say that f is locally conjugate to its linear part via an infinitely differentiable diffeomorphism? It turns out that (in general) a "non-resonance" condition must be satisfied for this to hold. The condition is the following: We assume that E is a finite dimensional space. Let $\lambda_1, \ldots, \lambda_n$ are the eigenvalues of A. The condition is

$$\lambda_i \neq \lambda_1^{m_1} \lambda_2^{m_2} \cdots \lambda_n^{m_n} \tag{8.10}$$

for any non-negative integers m_i with $\sum_i m_i \geq 2$. The C^∞ version of the theorem in my 1958 paper asserts that under this non-resonance condition f is indeed conjugate to its linear part via a C^∞ diffeomorphism.

8.2 Invariant manifolds.

Let p be a hyperbolic fixed point of a diffeomorphism, f. The *stable manifold* of f at p is defined as the set

$$W^s(p) = W^s(p, f) = \{x | \lim_{n \to \infty} f^n(x) = p\}. \tag{8.11}$$

Similarly, the *unstable manifold* of f at p is defined as

$$W^u(p) = W^u(p, f) = \{x | \lim_{n \to \infty} f^{-n}(x) = p\}. \tag{8.12}$$

We have defined W^s and W^u as *sets*. We shall see later on in this section that in fact they are submanifolds, of the same degree of smoothness as f. The terminology, while standard, is unfortunate. A point which is not exactly on $W^s(p)$ is swept away under iterates of f from any small neighborhood of p. This is the content of our first proposition below. So it is a very *unstable* property to lie on W^s. Better terminology would be "contracting" and "expanding" submanifolds. But the usage is standard, and we will abide by it. In any event, the sets $W^s(p)$ and $W^u(p)$ are, by their very definition, invariant under f.

In the case that $f = A$ is a hyperbolic *linear* transformation on a Banach space $E = S \oplus U$, then $W^s(0) = S$ and $W^u(0) = U$ as follows immediately from the definitions. The main result of this section will be to prove that in the general case, the stable manifold of f at p will be a submanifold whose tangent at p is the stable subspace of the linear transformation df_p.

Notice that for a hyperbolic fixed point, replacing f by f^{-1} interchanges the roles of W^s and W^u. So in much of what follows we will formulate and prove theorems for either W^s or for W^u. The corresponding results for W^u or for W^s then follow automatically.

More details in the linear case.

Let A be a hyperbolic linear transformation on a Banach space $E = S \oplus U$, and consider any ball, $B_r = B_r(0)$ of radius r about the origin. If $x \in B_r$ does *not* lie on $S \cap B_r$, this means that if we write $x = x_s \oplus x_u$ with $x_s \in S$ and $x_u \in U$ then $x_u \neq 0$. Then

$$\begin{aligned} \|A^n x\| &= \|A^n x_s\| + \|A^n x_u\| \\ &\geq \|A^n x_u\| \\ &\geq c^n \|x_u\|. \end{aligned}$$

If we choose n large enough, we will have $c^n\|x_u\| > r$. So eventually, $A^n x \notin B_r$. Put contrapositively,

$$S \cap B_r = \{x \in B_r | A^n x \in B_r \forall n \geq 0\}.$$

Back to the general case.

Now consider the case of a hyperbolic fixed point, p, of a diffeomorphism, f. We may introduce coordinates so that $p = 0$, and let us take $A = df_0$. By the C^0 conjugacy theorem, we can find a neighborhood, V of 0 and homeomorphism

$$h : B_r \to V$$

with

$$h \circ f = A \circ h.$$

Then

$$f^n(x) = h^{-1} \circ A^n \circ h\ (x)$$

will lie in U for all $n \geq 0$ if and only if $h(x) \in S(A)$ if and only if $A^n h(x) \to 0$. This last condition implies that $f^n(x) \to p$. We have thus proved:

Proposition 8.2.1. *Let p be a hyperbolic fixed point of a diffeomorphism, f. For any ball, $B_r(p)$ of radius r about p, let*

$$B_r^s(p) = \{x \in B_r(p) | f^n(x) \in B_r(p) \forall n \geq 0\}. \tag{8.13}$$

Then for sufficiently small r, we have

$$B_r^s(p) \subset W^s(p).$$

Furthermore, our proof shows that for sufficiently small r the set $B_r^s(p)$ is a topological submanifold in the sense that every point of $B_r^s(p)$ has a neighborhood (in $B_r^s(p)$) which is the image of a neighborhood, V in a Banach space under a homeomorphism, H. Indeed, the restriction of h to S gives the desired homeomorphism.

An important remark. In the general case we can not say that $B_r^s(p) = B_r(p) \cap W^s(p)$ because a point may escape from $B_r(p)$, wander around for a while, and then be drawn towards p.

But the proposition *does* assert that $B_r^s(p) \subset W^s(p)$ and hence, since W^s is invariant under f^{-1}, we have

$$f^{-n}[B_r^s(p)] \subset W^s(p)$$

for all n, and hence

$$\bigcup_{n \geq 0} f^{-n}[B_r^s(p)] \subset W^s(p).$$

On the other hand, if $x \in W^s(p)$, which means that $f^n(x) \to p$, eventually $f^n(x)$ arrives and stays in any neighborhood of p. Hence $p \in f^{-n}[B_r^s(p)]$ for some n. We have thus proved that for sufficiently small r we have

$$W^s(p) = \bigcup_{n \geq 0} f^{-n}[B_r^s(p)]. \tag{8.14}$$

We will prove that $B_r^s(p)$ is a submanifold. It will then follow from (8.14) that $W^s(p)$ is a submanifold. The global disposition of $W^s(p)$, and in particular its relation to the stable and unstable manifolds of other fixed points, is a key ingredient in the study of the long term behavior of dynamical systems. Her our focus is purely local, to prove something about the smooth character of the set $B_r^s(p)$. We follow the treatment in[**?**, Shub]

In fact, we will prove the local Lipschitzian character of the invariant manifolds, and refer to [Shub] for the proof of their smooth charater.

8.2.1 The Lipschitzian case.

We will begin with the hypothesis that f is merely Lipschitz, and give a proof (independent of the C^0 linearization theorem) of the existence and Lipschitz character of the W^u. We will work in the following situation: A is a hyperbolic linear isomorphism of a Banach space $E = S \oplus U$ with

$$\|Ax\| \leq a\|x\|, \ x \in S, \quad \|A^{-1}x\| \leq a\|x\|, \ \ x \in U,$$

where $0 < a < 1$.

We let $S(r)$ denote the ball of radius s about the origin in S, and $U(r)$ the ball of radius r in U.

We will assume that

$$f : S(r) \times U(r) \to E$$

is a Lipschitz map with

$$\|f(0)\| \leq \delta \tag{8.15}$$

and

$$\text{Lip}[f - A] \leq \epsilon. \tag{8.16}$$

We wish to prove:

Theorem 8.2.1. *Let $c < 1$. There exists an $\epsilon = \epsilon(a)$ and a $\delta = \delta(a,\epsilon, r)$ so that if f satisfies (8.15) and (8.16) then there is a map*

$$g : E_u(r) \to E_s(r)$$

with the following properties:
(i) g is Lipschitz with $\text{Lip}[g] \leq 1$.
(ii) The restriction of f^{-1} to graph(g) *is contracting and hence has a fixed point, p, on* graph(g).
(iii) We have

$$\text{graph}(g) = \bigcap f^n(S(r) \oplus U(r)) = W^u(p) \cap [S(r) \oplus U(p)].$$

The idea of the proof.

The idea of the proof is to apply the contraction fixed point theorem to the space of maps of $U(r)$ to $S(r)$. We want to identify such a map, v, with its graph (maybe we should write "cograph"):

$$\text{graph}(v) = \{(v(x), x),\ x \in U(r)\}.$$

Now

$$f[\text{graph}(v)] = \{f(v(x), x)\} = \{(f_s(v(x), x), f_u(v(x), x))\},$$

where we have introduced the notation

$$f_s = p_s \circ f, \quad f_u = p_u \circ f,$$

where p_s denotes projection onto S and p_u denotes projection onto U.

Suppose that the projection of $f[\text{graph}(v)]$ onto U is injective and its image contains $U(r)$. This means that for any $y \in U(r)$ there is a unique $x \in U(r)$ with

$$f_u(v(x), x) = y.$$

So we write

$$x = [f_u \circ (v, id)]^{-1}(y)$$

where we think of (v, id) as a map of $U(r) \to E$ and hence of

$$f_u \circ (v, id)$$

as a map of $U(r) \to U$. Then we can write

$$f[\text{graph}(v)] = \{(f_s(v([f_u \circ (v, id)]^{-1}(y), y)\} = [\text{graph}\, G_f(v)]$$

where

$$G_f(v) = f_s \circ (v, id) \circ [f_u \circ (v, id)]^{-1}. \tag{8.17}$$

The map $v \mapsto G_f(v)$ is called the **graph transform** (when it is defined). We are going to take

$$X = \text{Lip}_1(U(r), S(r))$$

to consist of all Lipschitz maps from $U(r)$ to $S(r)$ with Lipschitz constant ≤ 1. The purpose of the next few lemmas is to show that if ϵ and δ are sufficiently small then the graph transform, G_f is defined and is a contraction on X. The contraction fixed point theorem will then imply that there is a unique $g \in X$ which is fixed under G_f, and hence that graph(g) is invariant under f. We will then find that g has all the properties stated in the theorem.

Estimates on the graph transform.

In dealing with the graph transform it is convenient to use the box metric, $|\ |$, on $S \oplus U$ where

$$|x_s \oplus x_u| = \max\{\|x_s\|, \|x_u\|\}$$

i.e.

$$|x| = \max\{\|p_s(x)\|, \|p_u(x)\|\}.$$

This is equivalent, as a metric, to the original metric on E, by the definition of a direc sum decomposition.

We begin with

Lemma 8.2.1. *If $v \in X$ then*

$$\mathrm{Lip}[f_u \circ (v, id) - A_u] \leq \mathrm{Lip}[f - A].$$

Proof. Notice that $p_u \circ A(v(x), x) = p_u(A_s(v(x)), A_u x) = A_u x$ so

$$f_u \circ (v, id) - A_u = p_u \circ [f - A] \circ (v, id).$$

We have $\mathrm{Lip}[p_u] \leq 1$ since p_u is a projection, and

$$\mathrm{Lip}(v, id) \leq \max\{\mathrm{Lip}[v], \mathrm{Lip}[id]\} = 1$$

since we are using the box metric. □

Lemma 8.2.2. *Suppose that $0 < \epsilon < c^{-1}$ and*

$$\mathrm{Lip}[f - A] < \epsilon.$$

Then for any $v \in X$ the map $f_u \circ (v, id) : E_u(r) \to E_u$ is a homeomorphism whose inverse is a Lipschitz map with

$$\mathrm{Lip}\left[[f_u \circ (v, id)]^{-1}\right] \leq \frac{1}{c^{-1} - \epsilon}. \tag{8.18}$$

Proof. Using the preceding lemma, we have

$$\mathrm{Lip}[f_u - A_u] < \epsilon < c^{-1} < \|A_u^{-1}\|^{-1} = (\mathrm{Lip}[A_u])^{-1}.$$

By the Lipschitz implicit function theorem we conclude that $f_u \circ (v, id)$ is a homeomorphism with

$$\mathrm{Lip}\left[[f_u \circ (v, id)]^{-1}\right] \leq \frac{1}{\|A_u^{-1}\|^{-1} - \mathrm{Lip}[f_u \circ (v, id) - A_u]} \leq \frac{1}{c^{-1} - \epsilon}$$

by another application of the preceding lemma. □

We now wish to show that the image of $f_u \circ (v, id)$ contains $U(r)$ if ϵ and δ are sufficiently small: By the Lipschitz inverse function theorem, Theorem 6.5.1, we know that the image of $U(r)$ under $f_u \circ (v, id)$ contains a ball of radius r/λ about $[f_u \circ (v, id)](0)$ where λ is the Lipschitz constant of $[f_u \circ (v, id)]^{-1}$. By the preceding lemma, $r/\lambda = r(c^{-1} - \epsilon)$. Hence $f_u \circ (v, id)(U(r))$ contains the ball of radius

$$r(c^{-1} - \epsilon) - \|f_u(v(0), 0)\|$$

about the origin.

But

$$\begin{aligned}
\|f_u(v(0), 0)\| &\leq \|f_u(0,0)\| + \|f_u(v(0), 0) - f_u(0,0)\| \\
&\leq \|f_u(0,0)\| + \|(f_u - p_u A)(v(0), 0) - (f_u - p_u A)(0,0)\| \\
&\leq |f(0)| + |(f - A)(v(0), 0) - (f - A)(0,0)| \\
&\leq |f(0)| + \epsilon r.
\end{aligned}$$

The passage from the second line to the third is because $p_u A(x, y) = A_u y = 0$ if $y = 0$. The passage from the third line to the fourth is because we are using the box norm. So

$$r(c^{-1} - \epsilon) - \|f_u(v(0), 0)\| \geq r(c^{-1} - 2\epsilon) - \delta$$

if (8.15) holds. We would like this expression to be $\geq r$, which will happen if

$$\delta \leq r(c^{-1} - 1 - 2\epsilon). \tag{8.19}$$

We have thus proved

Proposition 8.2.2. *Let f be a Lipschitz map satisfying (8.15) and (8.16) where $2\epsilon < c^{-1} - 1$ and (8.19) holds. Then for every $v \in X$, the graph transform, $G_f(v)$ is defined and*

$$\mathrm{Lip}[G_f(v)] \leq \frac{c + \epsilon}{c^{-1} - \epsilon}.$$

The estimate on the Lipschitz constant comes from

$$\begin{aligned}
\mathrm{Lip}[G_f(v)] &\leq \mathrm{Lip}[f_s \circ (v, \mathrm{id})]\mathrm{Lip}[(f_u \circ (v, \mathrm{id})] \\
&\leq \mathrm{Lip}[f_s]\mathrm{Lip}[v]\mathrm{Lip} \cdot \frac{1}{c^{-1} - \epsilon} \\
&\leq (\mathrm{Lip}[A_s] + \mathrm{Lip}[p_s \circ (f - A)]) \cdot \frac{1}{c^{-1} - \epsilon} \\
&\leq \frac{c + \epsilon}{c^{-1} - \epsilon}.
\end{aligned}$$

In going from the first line to the second we have used the preceding lemma.

In particular, if

$$2\epsilon < c^{-1} - c \tag{8.20}$$

then

$$\mathrm{Lip}[G_f(v)] \leq 1.$$

Let us now obtain a condition on δ which will guarantee that

$$G_f(v)(U(r) \subset S(r).$$

Since

$$f_u \circ (v, \mathrm{id})U(r) \supset U(r),$$

we have

$$[f_u \circ (v, \mathrm{id})]^{-1}U(r) \subset U(r).$$

Hence, from the definition of $G_f(v)$, it is enough to arrange that

$$f_s \circ (v, \mathrm{id})[U(r)] \subset S(r).$$

For $x \in U(r)$ we have

$$\begin{aligned}
\|f_s(v(x), x)\| &\leq \|p_s \circ (f - A)(v(x), x)\| + \|A_s v(x)\| \\
&\leq |(f - A)(v(x), x)| + c\|v(x)\| \\
&\leq |(f - A)(v(x), x) - (f - A)(0, 0)| + |f(0)| + cr \\
&\leq \epsilon|(v(x), x)| + \delta + cr \\
&\leq \epsilon r + \delta + cr.
\end{aligned}$$

So we would like to have

$$(\epsilon + c)r + \delta < r$$

or

$$\delta \leq r(1 - c - \epsilon). \tag{8.21}$$

If this holds, then

$$G_f \textit{ maps } X \textit{ into } X.$$

When is $G_f : X \to X$ a contraction?

We now want conditions that guarantee that G_f is a contraction on X, where we take the sup norm. Let (w, x) be a point in $S(r) \oplus U(r)$ such that $f_u(w, x) \in U(r)$. Let $v \in X$, and consider

$$|(w, x) - (v(x), x)| = \|w - v(x)\|,$$

which we think of as the distance along S from the point (w, x) to graph(v). Suppose we apply f. So we replace (w, x) by $f(w, x) = (f_s(w, x), f_u(w, x))$ and graph(v) by $f(\mathrm{graph}(v)) = \mathrm{graph}(G_f(v))$. The corresponding distance along S is $\|f_s(w, x) - G_f(v)(f_u(w, x)\|$. We claim that

$$\|f_s(w, x) - G_f(v)(f_u(w, x))\| \leq (c + 2\epsilon)\|w - v(x)\|. \tag{8.22}$$

Indeed,

$$f_s(v(x), x) = G_f(v)(f_u(v(x), x)$$

by the definition of G_f, so we have

$$\begin{aligned}
\|f_s(w,x) - G_f(v)(f_u(w,x))\| &\leq \|f_s(w,x) - f_s(v(x),x)\| + \\
&\quad +\|G_f(v)(f_u((v(x),x) - G_f(v)(f_u(w,x))\| \\
&\leq \mathrm{Lip}[f_s]|(w,x) - (v(x),x)| + \\
&\quad +\mathrm{Lip}[f_u]|(v(x),x) - (w,x)| \\
&\leq \mathrm{Lip}[f_s - p_sA + p_sA]\|w - v(x)\| + \\
&\quad +\mathrm{Lip}[f_u - p_uA]\|w - v(x)\| \\
&\leq (\epsilon + c + \epsilon)\|w - v(x)\|
\end{aligned}$$

which is what was to be proved.

Consider two elements, v_1 and v_2 of X. Let z be any point of $U(r)$, and apply (8.22) to the point

$$(w,x) = (v_1([f_u \circ (v_1, \mathrm{id})]^{-1}](z)), [f_u \circ (v_1, \mathrm{id})]^{-1}](z))$$

which lies on graph(v_1), and where we take $v = v_2$ in (8.22). The image of (w,x) is the point $(G_f(v_1)(z), z)$ which lies on graph$(G_f(v_1))$, and, in particular, $f_u(w,x) = z$. So (8.22) gives

$$\|G_f(v_1)(z) - G_f(v_2)(z)\|$$

$$\leq (c + 2\epsilon)\|v_1([f_u \circ (v_1, \mathrm{id})]^{-1}](z)) - v_2([f_u \circ (v_1, \mathrm{id})]^{-1}](z)\|.$$

Taking the sup over z gives

$$\|G_f(v_1) - G_f(v_2)\|_{\mathrm{sup}} \leq (c + 2\epsilon)\|v_1 - v_2\|_{\mathrm{sup}}. \tag{8.23}$$

Intuitively, what (8.22) is saying is that G_f multiplies the S distance between two graphs by a factor of at most $(c + 2\epsilon)$. So G_f will be a contraction in the sup norm if

$$2\epsilon < 1 - c \tag{8.24}$$

which implies (8.20).

To summarize: we have proved that G_f is a contraction in the sup norm on X if (8.19), (8.21) and (8.24) hold, i.e.

$$2\epsilon < 1 - c, \quad \delta < r\min(c^{-1} - 1 - 2\epsilon,\ 1 - c - \epsilon).$$

Notice that since $c < 1$, we have $c^{-1} - 1 > 1 - c$ so both expressions occurring in the min for the estimate on δ are positive.

Now the uniform limit of continuous functions which all have $\mathrm{Lip}[v] \leq 1$ has Lipschitz constant ≤ 1. In other words, X is closed in the sup norm as a subset of the space of continuous maps of $U(r)$ into $S(r)$, and so we can apply the contraction fixed point theorem to conclude that there is a unique fixed point, $g \in X$ of G_f. Since $g \in X$, condition (i) of the theorem is satisfied.

As for (ii), let $(g(x), x)$ be a point on $graph(g)$ which is the image of the point $(g(y), y)$ under f, so

$$(g(x), x) = f(g(y), y)$$

which implies that

$$x = [f_u \circ (g, \mathrm{id})](y).$$

$$p_u \circ f_{|graph(g)} = [f_u \circ (g, \mathrm{id})] \circ (p_u)_{|graph(g)}.$$

In other words, the projection p_u conjugates the restriction of f to $graph(g)$ into $[f_u \circ (g, \mathrm{id})]$. Hence the restriction off^{-1} to $graph(g)$ is conjugated by p_u into $[f_u \circ (g, \mathrm{id})]^{-1}$. But, by (8.18), the map $[f_u \circ (g, \mathrm{id})]^{-1}$ is a contraction since

$$c^{-1} - 1 > 1 - c > 2\epsilon$$

so

$$c^{-1} - \epsilon > 1 + \epsilon > 1.$$

The fact that $\mathrm{Lip}[g] \leq 1$ implies that

$$|(g(x), x) - (g(y), y)| = \|x - y\|$$

since we are using the box norm. So the restriction of p_u to $graph(g)$ is an isometry between the (restriction of) the box norm on $graph(g)$and the norm on U. So we have proved statement (ii), that the restriction of f^{-1} to $graph(g)$ is a contraction.

We now turn to statement (iii) of the theorem. Suppose that (w, x) is a point in $S(r) \oplus U(r)$ with $f(w, x) \in S(r) \oplus U(r)$. By (8.22) we have

$$\|f_s(w, x) - g(f_u(w, x)\| \leq (c + 2\epsilon)\|w - g(x)\|$$

since $G_f(g) = g$. So if the first n iterates of f applied to (w, x) all lie in $S(r) \oplus U(r)$, and if we write

$$f^n(w, x) = (z, y),$$

we have

$$\|z - g(y)\| \leq (c + 2\epsilon)^n\|w - g(x)\| \leq (c + 2\epsilon)r.$$

So if the point (z, y) is in $\bigcap f^n(S(r) \oplus U(r))$ we must have $z = g(y)$, in other words

$$\bigcap f^n(S(r) \oplus U(r)) \subset \mathrm{graph}(g).$$

But

$$\mathrm{graph}(g) = f[\mathrm{graph}(g)] \cap [S(r) \oplus U(r)]$$

so

$$\mathrm{graph}(g) \subset \bigcap f^n(S(r) \oplus U(r)),$$

proving that

$$\text{graph}(g) = \bigcap f^n(S(r) \oplus U(r)).$$

We have already seen that the restriction of f^{-1} to graph(g) is a contraction, so all points on graph(g) converge under the iteration of f^{-1} to the fixed point, p. So they belong to $W^u(p)$. This completes the proof of the theorem. □

Notice that if $f(0) = 0$, then $p = 0$ is the unique fixed point.

Chapter 9

The Perron-Frobenius theorem.

The theorem we will discuss in this chapter (to be stated below) about matrices with non-negative entries, was proved, for matrices with strictly positive entries, by Oskar Perron (1880-1975) in 1907 and extended by Ferdinand Georg Frobenius (1849-1917) to matrices which have non-negative entries and are irreducible (definition below) in 1912.

This theorem has miriads of applications, several of which we will study in this book.

9.1 Non-negative and positive matrices.

We begin with some definitions.

We say that a real matrix T is **non-negative** (or **positive**) if all the entries of T are non-negative (or positive). We write $T \geq 0$ or $T > 0$. We will use these definitions primarily for square ($n \times n$) matrices and for column vectors $= (n \times 1)$ matrices, although rectangular matrices will come into the picture at one point.

The positive orthant.

We let

$$Q := \{x \in \mathbb{R}^n : x \geq 0, \quad x \neq 0\}$$

so Q is the non-negative orthant excluding the origin, which(by abuse of language) we will call the **positive orthant** . Also let

$$C := \{x \geq 0 : \|x\| = 1\}.$$

So C is the intersection of the positive orthant with the unit sphere.

9.1.1 Primitive and irreducible non-negative square matrices.

A non-negative matrix square T is called **primitive** if there is a k such that all the entries of T^k are positive. It is called **irreducible** if for any i, j there is a $k = k(i, j)$ such that $(T^k)_{ij} > 0$.

If T is irreducible then $I + T$ is primitive. Indeed, the binomial expansion

$$(I+T)^k = I + kT + \frac{k(k-1)}{2}T^2 + \cdots$$

will eventually have positive entries in all positions if k large enough.

9.1.2 Statement of the Perron-Frobenius theorem.

In the statement of the Perron-Frobenius theorem we assume that T is irreducible. We now state the theorem:

Theorem 9.1.1. *Let T be an irreducible matrix.*

1. *T has a positive (real) eigenvalue $\lambda_{\max}$ such that all other eigenvalues of T satisfy*

$$|\lambda| \leq \lambda_{\max}.$$

2. *Furthermore $\lambda_{\max}$ has algebraic and geometric multiplicity one, and has an eigenvector x with $x > 0$.*

3. *Any non-negative eigenvector is a multiple of x.*

4. *More generally, if $y \geq 0$, $y \neq 0$ is a vector and μ is a number such that*

$$Ty \leq \mu y$$

then

$$y > 0, \quad \textit{and} \quad \mu \geq \lambda_{\max}$$

with $\mu = \lambda_{\max}$ if and only if y is a multiple of x.

5. *If $0 \leq S \leq T$, $S \neq T$ then every eigenvalue σ of S satisfies*

$$|\sigma| < \lambda_{\max}.$$

6. *In particular, all the diagonal minors $T_{(i)}$ obtained from T by deleting the i-th row and column have eigenvalues all of which have absolute value $< \lambda_{\max}$.*

7. *If T is primitive, then all other eigenvalues of T satisfy*

$$|\lambda| < \lambda_{\max}.$$

9.1.3 Proof of the Perron-Frobenius theorem.

We now embark on the proof of this important theorem.

Let

$$P := (I + T)^k$$

where k is chosen so large that P is a positive matrix. Then $v \leq w, v \neq w \Rightarrow Pv < Pw$.

Recall that Q denotes the positive orthant and that C denotes the intersection of the unit sphere with the positive orthant. For any $z \in Q$ let

$$L(z) := \max\{s : sz \leq Tz\} = \min_{1 \leq i \leq n, z_i \neq 0} \frac{(Tz)_i}{z_i}. \tag{9.1}$$

By definition $L(rz) = L(z)$ for any $r > 0$, so $L(z)$ depends only on the ray through z. If $z \leq y$, $z \neq y$ we have $Pz < Py$. Also $PT = TP$. So if $sz \leq Tz$ then

$$sPz \leq PTz = TPz$$

so

$$L(Pz) \geq L(z).$$

Furthermore, if $L(z)z \neq Tz$ then $L(z)Pz < TPz$. So $L(Pz) > L(z)$ unless z is an eigenvector of T with eigenvalue $L(z)$.

This suggests a plan for the proof: that we look for a positive vector which maximizes L, show that it is the eigenvector we want in the theorem and establish the properties stated in the theorem.

Finding a positive eigenvector.

Consider the image of C under P. It is compact (being the image of a compact set under a continuous map) and all of the elements of $P(C)$ have all their components strictly positive (since P is positive). Hence the function L is continuous on $P(C)$. Thus L achieves a maximum value, $L_{\max}$ on $P(C)$. Since $L(z) \leq L(Pz)$ this is in fact the maximum value of L on all of Q, and since $L(Pz) > L(z)$ unless z is an eigenvector of T, we conclude that

$L_{\max}$ is achieved at an eigenvector, call it x of T and $x > 0$ with $L_{\max}$ the eigenvalue.

Since $Tx > 0$ and $Tx = L_{\max}x$ we have $L_{\max} > 0$.

Showing that $L_{\max}$ is the maximum eigenvalue.

Let y be any eigenvector with eigenvalue λ, and let $|y|$ denote the vector whose components are $|y_j|$, the absolute values of the components of y. We have $|y| \in Q$ and from

$$Ty = \lambda y$$

which says that

$$\lambda y_i = \sum_j T_{ij} y_j$$

and the fact that the $T_{ij} \geq 0$ we conclude that

$$|\lambda| y_i| \leq \sum_i T_{ij} |y_j|$$

which we write for short as

$$|\lambda||y| \leq T|y|.$$

Recalling the definition (9.1) of L, this says that $|\lambda| \leq L(|y|) \leq L_{\max}$. So we may use the notation

$$\lambda_{\max} := L_{\max}$$

since we have proved that

$$|\lambda| \leq \lambda_{\max}.$$

We have proved item 1 in the theorem.

Notice that we can not have $\lambda_{\max} = 0$ since then T would have all eigenvalues zero, and hence be nilpotent, contrary to the assumption that T is irreducible. So

$$\lambda_{\max} > 0.$$

Showing that $0 \leq S \leq T,\ S \neq T \Rightarrow \lambda_{\max}(S) \leq \lambda_{\max}(T)$.

Suppose that $0 \leq S \leq T$. If $z \in Q$ is a vector such that $sz \leq Sz$ then since $Sz \leq Tz$ we get $sz \leq Tz$ so $L_S(z) \leq L_T(z)$ for all z and hence

$$0 \leq S \leq T \quad \Rightarrow \quad L_{\max}(S) \leq L_{\max}(T).$$

So

$$0 \leq S \leq T,\ S \neq T \Rightarrow \lambda_{\max}(S) \leq \lambda_{\max}(T)$$

Showing that $\lambda_{\max}(T^\dagger) = \lambda_{\max}(T)$.

We may apply the previous results to $T^\dagger$, the transpose of T, to conclude that it also has a positive maximum eigenvalue. Let us call it η. (We shall soon show that $\eta = \lambda_{\max}$.) This means that there is a row vector $w > 0$ such that

$$w^\dagger T = \eta w^\dagger.$$

Recall that $x > 0$ denotes the eigenvector with maximum eigenvalue $\lambda_{\max}$ of T. We have

$$w^\dagger T x = \eta w^\dagger x = \lambda_{\max} w^\dagger x$$

implying that $\eta = \lambda_{\max}$ since $w^\dagger x > 0$.

Proving the first two assertions in item 4 of the theorem.

Suppose that $y \in Q$ and $Ty \leq \mu y$. Then

$$\lambda_{\max} w^\dagger y = w^\dagger T y \leq \mu w^\dagger y$$

implying that $\lambda_{\max} \leq \mu$, again using the fact that all the components of w are positive and some component of y is positive so $w^\dagger y > 0$. In particular, if $Ty = \mu y$ then then $\mu = \lambda_{\max}$.

Furthermore, if $y \in Q$ and $Ty \leq \mu y$ then $\mu \geq 0$ and

$$0 < Py = (I+T)^{n-1} y \leq (1+\mu)^{n-1} y$$

so

$$y > 0.$$

This proves the first two assertions in item 4.

If $\mu = \lambda_{\max}$ then $w^\dagger(Ty - \lambda_{\max} y) = 0$ but $Ty - \lambda_{\max} y \leq 0$ and therefore $w^\dagger(Ty - \lambda_{\max} y) = 0$ implies that $Ty = \lambda_{\max} y$. Then the last assertion of item 4) - that y is a scalar multiple of x - will then follow from item 2) - that $\lambda_{\max}$ has multiplicity one - once we prove item 2), since we have shown that y must be an eigenvector with eigenvalue $\lambda_{\max}$.

Proof that if $0 \leq S \leq T$, $S \neq T$ then every eigenvalue σ of S satisfies $|\sigma| < \lambda_{\max}$.

Suppose that $0 \leq S \leq T$ and $Sz = \sigma z$, $z \neq 0$. Then

$$T|z| \geq S|z| \geq |\sigma||z|$$

so

$$|\sigma| \leq L_{\max}(T) = \lambda_{\max},$$

as we have already seen. But if $|\sigma| = \lambda_{\max}(T)$ then $L_T(|z|) = L_{\max}(T)$ so $|z| > 0$ and $|z|$ is also an eigenvector of T with the same eigenvalue. But then $(T - S)|z| = 0$ and this is impossible unless $S = T$ since $|z| > 0$.

Replacing the i-th row and column of T by zeros give an $S \geq 0$ with $S < T$ since the irreducibility of T precludes all the entries in a row being. This proves the assertion that the eigenvalues of T_i are all less in absolute value that $\lambda_{\max}$. zero.

A lemma in linear algebra.

Let T be a (square) matrix and let Λ be a diagonal matrix of the same size, with entries $\lambda_1, \ldots, \lambda_n$ along the diagonal. Expanding $\det(\Lambda - T)$ along the i-th row shows that

$$\frac{\partial}{\partial \lambda_i} \det(\Lambda - T) = \det(\Lambda_i - T_i)$$

where the subscript i means the matrix obtained by eliminating the i-th row and the i-th column from each matrix.

Setting $\lambda_i = \lambda$ and applying the chain rule from calculus, we get

$$\frac{d}{d\lambda}\det(\lambda I - T) = \sum_i \det(\lambda I - T_{(i)})$$

So from linear algebra we know that

$$\frac{d}{d\lambda}\det(\lambda I - T) = \sum_i \det(\lambda I - T_{(i)}).$$

Showing that $\lambda_{\max}$ has algebraic (and hence geometric) multiplicity one.

Each of the matrices $\lambda_{\max} I - T_{(i)}$ has Each of the matrices $\lambda_{\max} I - T_{(i)}$ has strictly positive determinant by what we have just proved. This shows that the derivative of the characteristic polynomial of T is not zero at $\lambda_{\max}$, and therefore the algebraic multiplicity and hence the geometric multiplicity of $\lambda_{\max}$ is one. This proves 2) and hence all but the last assertion of the theorem, which says that if T is primitive, then all the other eigenvalues of T satisfy

$$|\lambda| < \lambda_{\max}.$$

Proof of the last assertion of the theorem.

The eigenvalues of T^k are the k-th powers of the eigenvalues of T. So if we want to show that there are no other eigenvalues of a primitive matrix with absolute value equal to $\lambda_{\max}$, it is enough to prove this for a positive matrix. Dividing the positive matrix by $\lambda_{\max}$, we are reduced to proving the following

Lemma 9.1.1. *Let $A > 0$ be a positive matrix with $\lambda_{\max} = 1$. Then all other eigenvalues of A satisfy $|\lambda| < 1$.*

Proof of the lemma. Suppose that z is an eigenvector of A with eigenvalue λ with $|\lambda| = 1$. Then $|z| = |\lambda z| = |Az| \leq| A||z| = A|z| \Rightarrow| z| \leq A|z|$. Let $y := A|z| -| z|$ so $y \geq 0$. Suppose (contrary to fact) that $y \neq 0$. Then $Ay > 0$ and $A|z| > 0$ so there is an $\epsilon > 0$ so that $Ay > \epsilon A\ |z|$ and hence $A(A|z| -| z|) > \epsilon A\ |z|$ or

$$B(A|z|) > A|z|, \quad \text{where } B := \frac{1}{1+\epsilon}A.$$

This implies that $B^k A|z| > A|z|$ for all k. But the eigenvalues of B are all < 1 in absolute value, so $B^k \to 0$. Thus all the entries of $A|z|$ are ≤ 0 contradicting the fact that $A|z| > 0$. So $|z|$ is an eigenvector of A with eigenvalue 1.

But $|Az| = |z|$ so $|Az| = A|z|$ which can only happen if all the entries of z are of the same sign. So z must be a multiple of our eigenvector x since there are no other eigenvectors with all entries of the same sign other than multiples of x So $\lambda = 1$. □

This completes the proof of the theorem. We still must discuss what happens in the non-primitive irreducible case. We will find that there is a nice description also due to Frobenius. But first some examples:

Examples for two by two matrices.

To check whether a matrix with non-negative entries is primitive, or irreducible, or neither, we may replace all of the non-zero entries by ones since this does not affect the classification. The matrix

$$\begin{pmatrix} 1 & 1 \\ 1 & 1 \end{pmatrix}$$

is (strictly) positive hence primitive. The matrices

$$\begin{pmatrix} 1 & 0 \\ 1 & 1 \end{pmatrix} \quad \text{and} \quad \begin{pmatrix} 1 & 1 \\ 0 & 1 \end{pmatrix}$$

both have 1 as a double eigenvalue so can not be irreducible.

The matrix $\begin{pmatrix} 1 & 1 \\ 1 & 0 \end{pmatrix}$ satisfies

$$\begin{pmatrix} 1 & 1 \\ 1 & 0 \end{pmatrix}^2 = \begin{pmatrix} 2 & 1 \\ 1 & 1 \end{pmatrix}$$

and so is primitive. Similarly for $\begin{pmatrix} 0 & 1 \\ 1 & 1 \end{pmatrix}$.

The matrix $\begin{pmatrix} 0 & 1 \\ 1 & 0 \end{pmatrix}$ is irreducible but not primitive. Its eigenvalues are 1 and -1.

9.2 Graphology.

9.2.1 Non-negative matrices and directed graphs.

A **directed graph** is a pair consisting of a set V (called **vertices** or **nodes**) and a subset $E \subset V \times V$ called (directed) **edges**. The directed edge (v_i, v_j) "goes from v_i to v_j. We draw it as an arrow.

The **graph associated to the non-negative square matrix** M of size $n \times n$ has $V = \{v_1, \dots, v_n\}$ and the directed edge

$$(v_j, v_i) \in E \iff M_{ij} \neq 0.$$

(Notice the reversal of order in this convention. Sometimes the opposite convention is used.)

The **adjacency matrix** A of the graph (V, E) is the $n \times n$ matrix (where n is the number of nodes) with $A_{ij} = 1$ if $(v_j, v_i) \in E$ and $= 0$ otherwise.

So if (V, E) is associated to M and A is its adjacency matrix, then A is obtained from M by replacing its non-zero entries by ones.

Paths and powers.

A **path** from a vertex v to a vertex w is a finite sequence $v_0, \ldots, v_\ell$ with $v_0 = v$, $v_\ell = w$ where each (v_i, v_{i+1}) is an edge. The number ℓ, i,e, the number of edges in the path is called the **length** of the path.

If A is the adjacency matrix of the graph, then $(A^2)_{ij}$ gives the number of paths of length two joining v_j to v_i, and, more generally, $(A^\ell)_{ij}$ gives the number of paths of length ℓ joining v_j to v_i.

So M is irreducible $\iff$ its associated graph is **strongly connected** in the sense that for any two vertices v_i and v_j there is a path (of some length) joining v_i to v_j.

What is a graph theoretical description of primitivity? We now discuss this question.

9.2.2 Cycles and primitivity.

A **cycle** is a path starting and ending at the same vertex.

Let M be primitive with, say M^k strictly positive. Then the associated graph is strongly connected, indeed every vertex can be joined to every other vertex by a path of length k. But then every vertex can be joined to itself by a path of length k, so there are (many) cycles of length k.

But then M^{k+1} is also strictly postive and hence there are cycles of length $k+1$. So there are (at least) two cycles whose lengths are relatively prime.

We will now embark on proving the converse:

Theorem 9.2.1. *If the graph associated to M is strongly connected and has two cycles of relatively prime lengths, then M is primitive.*

We will use the following elementary fact from number theory whose proof we will give after using it to prove the theorem:

Lemma 9.2.1. *Let a and b be positive integers with g.c.d.$(a, b) = 1$. Then there is an integer B such that every integer $\geq B$ can be written as an integer combination of a and b with non-negative coefficients.*

We will prove the theorem from the lemma by showing that for

$$k := 3(n-1) + B$$

there is a path of length k joining any pair of vertices.

We can construct a path going from v to w by going from v to a point x on the first cycle, going around this cycle a number of times, then joining x to a point y on the second cycle, going around this cycle a number of times, and then going from y to w.

The paths from v to x, from x to y, from y to w have total lengths at most $3(n-1)$. But then, by the lemma, we can make up the difference between this total length and k by going around the cycles an appropriate number of times. □

Proof of the lemma. An integer n can be written as $ia + jb$ with i and j non-negative integers $\iff$ it is in one of the following sequences

$$\begin{array}{llll} 0, & b, & 2b, & \ldots, \\ a, & b+a, & 2b+a & \ldots \\ \vdots & & & \\ (b-1)a, & b+(b-1)a, & 2b+(b-1)a, & \ldots \end{array}.$$

Since a and b are relatively prime, the elements of the first column all belong to different conjugacy classes mod b, So every integer n can be written as $n = ra+sb$ where $0 \leq r < b$. If $s < 0$ then $n < a(b-1)$. □

A mild extension of the above argument will show that if there are several (not necessarily two) cycles whose greatest common denominator is one, then M is primitive.

9.2.3 The Frobenius analysis of the irreducible non-primitive case.

In this section I follow the exposition of Mike Boyle "NOTES ON THE PERRON-FROBENIUS THEORY OF NONNEGATIVE MATRICES " available on the web.

The definition of the period on an irreducible matrix.

The **period** of an irreducible non-negative matrix A is the greatest common divisor of the lengths of the cycles in the associated graph.

The Frobenius form of an irreducible non-primitive matrix.

Let A be an irreducible non-negative matrix A with period $p > 1$. Let v be any vertex in the associated graph. For $0 \leq i < p$ let

$$C_i := \{u| \text{ there is a path of length } n \text{ from } u \text{ to } v \text{ with } n \equiv i \bmod p\}.$$

Since A is irreducible, every vertex belongs to one of the sets C_i, and by the definition of p, it can belong to only one. So the sets C_i partition the vertex set. Let us relabel the vertices so that the first $\#(C_0)$ vertices belong to C_0, the second $\#(C_1)$ vertices belong to C_1 etc. This means that we have permutation of the integers P so that PAP^{-1} has a block form with rectangular blocks which looks something like a cyclic permutation matrix. For example, for $p = 4$, the matrix PAP^{-1} would look like

$$\begin{pmatrix} 0 & A_1 & 0 & 0 \\ 0 & 0 & A_2 & 0 \\ 0 & 0 & 0 & A_3 \\ A_4 & 0 & 0 & 0 \end{pmatrix}.$$

I want to emphasize that the matrices A_i are rectangular, not necessarily square.

The eigenvalues of an irreducible non-primitive matrix.

Since the spectral properties of PAP^{-1} and A are the same, we will assume from now on that A is in the block form. To illustrate the next step in Frobenius's analysis, let us go back to the $p = 4$ example, and raise A to the fourth power, and obtain a block diagonal matrix:

$$\begin{pmatrix} 0 & A_1 & 0 & 0 \\ 0 & 0 & A_2 & 0 \\ 0 & 0 & 0 & A_3 \\ A_4 & 0 & 0 & 0 \end{pmatrix}^4$$

$$= \begin{pmatrix} A_1A_2A_3A_4 & 0 & 0 & 0 \\ 0 & A_2A_3A_4A_1 & 0 & 0 \\ 0 & 0 & A_3A_4A_1A_2 & 0 \\ 0 & 0 & 0 & A_4A_1A_2A_3 \end{pmatrix}.$$

Each of these diagonal blocks has period one and so is primitive. Also, if $D(i)$ denotes the i-th diagonal block, then there are rectangular matrices R and S such that

$$D(i) = SR \quad \text{and} \quad D(i+1) = RS.$$

If we take $i = 2$ in the above example, $S = A_2$ and $R = A_3A_4A_1$.

Therefore, taking their k-th power, we have

$$D(i)^k = S(RS)^{k-1}R, \quad \text{and} \quad D(i+1)^k = ((RS)^{k-1}R)S.$$

This implies that $D(i)^k$ and $D(i+1)^k$ have the same trace. Since the trace of the k-th power of a matrix is the sum of the k-th power of its eigenvalues, we conclude that the non-zero eigenvalues of each of the $D(i)$ are the same.

Proposition 9.2.1. *Let A be a non-negative irreducible matrix with period p and let ω be a primitive p-th root of unity, for example $\omega = e^{2\pi i/p}$. Then the matrices A and ωA are conjugate. In particular, if c is an eigenvalue of A with multiplicity m so is ωc.*

The following computation for $p = 3$ explains the general case:

$$\begin{pmatrix} \omega^{-1}I & 0 & 0 \\ 0 & \omega^{-2}I & 0 \\ 0 & 0 & I \end{pmatrix} \begin{pmatrix} 0 & A_1 & 0 \\ 0 & 0 & A_2 \\ A_3 & 0 & 0 \end{pmatrix} \begin{pmatrix} \omega I & 0 & 0 \\ 0 & \omega^2 I & 0 \\ 0 & 0 & I \end{pmatrix}$$

$$= \begin{pmatrix} 0 & \omega A_1 & 0 \\ 0 & 0 & \omega A_2 \\ \omega A_3 & 0 & 0 \end{pmatrix} = \omega \begin{pmatrix} 0 & A_1 & 0 \\ 0 & 0 & A_2 \\ A_3 & 0 & 0 \end{pmatrix}. \qquad \square$$

A supplement to the Perron-Frobenius theorem.

So we can supplement the Perron-Frobenius theorem in the case that A is a non-negative irreducible matrix of period p by

Theorem 9.2.2. *Let A be a non-negative irreducible matrix of period p with maximum real eigenvalue $\lambda_{\max}$. The eigenvalues λ of A with $|\lambda| = \lambda_{\max}$ are all simple and of the form $\omega\lambda_{\max}$ as ω ranges over the p-th roots of unity.*

The spectrum of A is invariant under multiplication by ω where ω is a primitive p-th root of unity.

9.3 Asymptotic behavior of powers of a primitive matrix.

Let A be a primitive matrix and r its maximal eigenvalue as given by the Perron-Frobenius theorem. Let $x > 0$ be a (right-handed) eigenvector of A with eigenvalue r, so $Ax = rx$ and we choose x so that $x > 0$. Let $y > 0$ be a (row) vector with $yA = ry$ (also determined up to scalar multiple by a positive number and let us choose y so that $y \cdot x = 1$.

The rank one matrix $H := x \otimes y^\dagger$ has image space R, the one dimensional space spanned by x and

$$H^2 = H$$

so H is a projection. The operator $I - H$ is then also a projection whose image is the null space N of H. Also $AH = Ax \otimes y = rx \otimes y = x \otimes ry = HA$. So we have the direct sum decomposition of our space as $R \oplus N$ which is invariant under A. We have the direct sum decomposition of our space as $R \oplus N$ which is invariant under A.

The restriction of A to N has all its eigenvalues strictly less than r in absolute value, while the restriction of A to the one dimensional space R is multiplication by r. So if we set

$$P := \frac{1}{r}A$$

then the restriction of P to N has all its eigenvalues < 1 in absolute value. The above decomposition is invariant under all powers of P and the restriction of

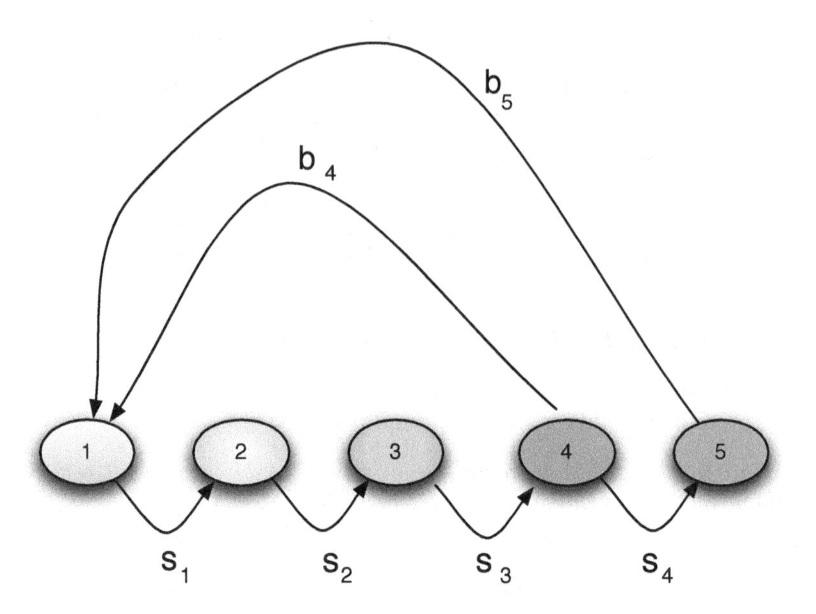

Figure 9.1: 5 age groups, the last two child bearing.

P^k to N tends to zero as $k \to \infty$, while the restriction of P to R is the identity. So we have proved

Theorem 9.3.1.

$$\lim_{k\to\infty}\left(\frac{1}{r}A\right)^k = H.$$

We now turn to a varied collection of applications of the preceding result.

9.4 The Leslie model of population growth.

In 1945 Leslie introduced a model for the growth of a stratified population: The population to consider consists of the females of a species, and the stratification is by age group. (For example into females under age 5, between 5 and 10, between 10 and 15 etc.) So the population is described by a vector whose size is the number of age groups and whose i-th component is the number of females in the i-th age group.

He let b_i be the expected number of daughters produced by a female in the i-th age group and s_i the proportion of females in the i-th age group who survive (to the next age group) in one time unit.

The Leslie matrix.

So the transition after one time unit is given by the **Leslie matrix**

$$L = \begin{pmatrix} b_1 & b_2 & \cdots & b_{n-1} & b_n \\ s_1 & 0 & \cdots & \cdots & 0 \\ 0 & s_2 & 0 & \cdots & 0 \\ \vdots & \vdots & \ddots & \ddots & \vdots \\ 0 & 0 & \cdots & s_{n-1} & 0 \end{pmatrix}.$$

In this matrix we might as well take $b_n > 0$ as there is no point in taking into consideration those females who are past the age of reproduction as far as the long term behavior of the populaton is concerned. Also we restrict ourselves to the case where all the $s_i > 0$ since otherwise the population past age i will die out.

The Leslie matrix is irreducible.

The graph associated to L consists of n vertices with $v_1 \to v_2 \to \cdots \to v_n$ with v_n (and possibly others) connected to v_1 and so is strongly connected. So L is irreducible.

What is the positive eigenvector?

We might as well take the first component of the positive eigenvector to be 1. The elements in the second to the last positions in Lx are then determined recursively by

$$x_2 = s_1,\ x_3 = s_2 x_2, \ldots.$$

Then the equation $Lx = rx$ tells us that

$$x_2 = \frac{s_1}{r},\ x_3 = \frac{s_1 s_2}{r^2}, \cdots$$

and then the first component of $Lx = rx$ tell us that r is a solution to the equation

$$p(r) = 1$$

where

$$p(r) = \frac{b_1}{r} + \frac{b_2 s_1}{r^2} + \cdots + \frac{b_n s_1 \cdots s_{n-1}}{r^n}.$$

The function $p(r)$ is defined for $r > 0$, is strictly decreasing, tends to ∞ as $r \to 0$ and to 0 as $r \to \infty$ and so the equation $p(r) = 1$ has a unique positive root as we expect from the general theory.

9.4.1 When is the Leslie matrix primitive?

Each i with $b_i > 0$ gives rise to a cycle of length i in the graph. So if there are two i-s with $b_i > 0$ which are relatively prime to one another then L is primitive. (In fact, as mentioned above, an examination of the proof of the corresponding fact in the general Perron-Frobenius theorem shows that it is enough to know that there are i's whose greatest common divisor is 1 with $b_i > 0$.) In particular, if $b_i > 0$ and $b_{i+1} > 0$ for some i then L is primitive.

9.4.2 The limiting behavior when the Leslie matrix is primitive.

If L is primitive with maximal eigenvalue r then we know from the general Perron Frobenius theory that the total population grows (or declines) approximate the rate r^k and that the relative size of the age groups to the general population is proportional to the positive eigenvector (as computed above).

Fibonacci.

The most famous and (ancient) Leslie matrix is the two by two matrix

$$F = \begin{pmatrix} 1 & 1 \\ 1 & 0 \end{pmatrix}$$

whose powers when applied to $\begin{pmatrix} 1 \\ 0 \end{pmatrix}$ generate the Fibonacci numbers. The eigenvalues of F are

$$\frac{1 \pm \sqrt{5}}{2}.$$

An imprimitive Leslie matrix.

If the females give birth only in the last time period then the Leslie matrix is not primitive. For example, Atlantic salmon die immediately after spawning. Assuming, for example, that there are three age groups, we obtain the Leslie matrix

$$L = \begin{pmatrix} 0 & 0 & b \\ s_1 & 0 & 0 \\ 0 & s_2 & 0 \end{pmatrix}.$$

The characteristic polynomial of this matrix is

$$\lambda^3 - bs_1s_2$$

so if F is the real root of $F^3 = bs_1s_2$ the eigenvalues are

$$F, \omega F, \omega^2 F$$

where ω is a primitive cube root of unity. So L is conjugate to the matrix

$$\begin{pmatrix} 0 & 0 & F \\ F & 0 & 0 \\ 0 & F & 0 \end{pmatrix} = \begin{pmatrix} F & 0 & 0 \\ 0 & F & 0 \\ 0 & 0 & F \end{pmatrix} \begin{pmatrix} 0 & 0 & 1 \\ 1 & 0 & 0 \\ 0 & 1 & 0 \end{pmatrix}.$$

So

$$\frac{1}{F} L$$

is periodic with period 3.

For a thorough discussion of the implementation of the Leslie model, see the book [?].

9.5 Markov chains in a nutshell.

A non-negative matrix M is a **stochastic** matrix if each of the row sums equal 1. Then the column vector $\mathbf{1}$ all of whose entries equal 1 is an eigenvector with eigenvalue 1. So if M is irreducible 1 is the maximal eigenvalue since $\mathbf{1}$ has all positive entries.

If M is primitive, then we know from the general theory that

$$M^k \to \begin{pmatrix} \pi_1 & \pi_2 & \cdots & \pi_n \\ \pi_1 & \pi_2 & \cdots & \pi_n \\ \vdots & \vdots & \vdots & \vdots \\ \pi_1 & \pi_2 & \cdots & \pi_n \end{pmatrix}$$

where $\mathbf{p} := (\pi_1, \pi_2, \cdots, \pi_n)$ is the unique vector whose entries sum to one and satisfies $\mathbf{p}M = \mathbf{p}$.

9.6 The Google ranking.

In this section, we follow the discussion in Chapters 3 and 4 of [Langville and Meyer]

The issue is how to rank the "importance" of URL's on the web. The idea is to think of a hyperlink from A to B as an endorsement of B. So many inlinks should increase the value of a URL. On the other hand, each inlink should carry a weight. A recommendation should carry more weight if coming from an important source, but less if the source is known to have many outlinks. (If I am known to write many positive letters of recommendation then the value of each decreases, even though I might be an "important" professor.)

9.6.1 The basic equation.

So we would like the ranking to satisfy an equation like

$$r(P_i) = \sum_{P_j \in B_{P_i}} \frac{r(P_j)}{|P_j|} \tag{9.2}$$

where $r(P)$ is the desired ranking, B_{P_i} is the set of "pages" pointing into P_i, and $|P_j|$ is the number of links pointing out of P_j.

The matrix H.

So if $\mathbf{r}$ denotes the row vector whose i-th entry is $r(P_i)$ and $\mathbf{H}$ denotes the matrix whose ij entry is $1/|P_i|$ if there is a link from P_i to P_j then (9.2) becomes

$$\mathbf{r} = \mathbf{r}\mathbf{H}. \tag{9.3}$$

The matrix $\mathbf{H}$ is of size $n \times n$ where n is the number of "pages", roughly 12 billion of so at the current time. We would like to solve the above equation by iteration, as in the case of a Markov chain. Despite the huge size, computing products with $\mathbf{H}$ is feasible because $\mathbf{H}$ is sparse, i.e. it consists mostly of zeros.

9.6.2 Problems with H, the matrix S.

The matrix $\mathbf{H}$ will have some rows consisting entirely of zeros. These correspond to the "dangling nodes", pages (such as pdf. files etc.) which have no outgoing links. Other than these, the row sums are one.

To fix this problem, Brin and Page, the inventors of Google, replaced the zero rows by rows consisting entirely of $1/n$ (a very small number). So let $\mathbf{a}$ denote the column vector whose i-th entry is 1 if the i-th row is dangling row, and $a_i = 0$ otherwise. Let $\mathbf{e}$ be the row vector consisting entirely of ones. Brin and Page replace $\mathbf{H}$ by

$$\mathbf{S} := \mathbf{H} + \frac{1}{n}\mathbf{a} \otimes \mathbf{e}.$$

The matrix $\mathbf{S}$ is now a Markov chain matrix, all rows sum to one.

For example, suppose that node 2 is a dangling mode and that the matrix $\mathbf{H}$ is

$$\mathbf{H} = \begin{pmatrix} 0 & \frac{1}{2} & \frac{1}{2} & 0 & 0 & 0 \\ 0 & 0 & 0 & 0 & 0 & 0 \\ \frac{1}{3} & \frac{1}{3} & 0 & 0 & \frac{1}{3} & 0 \\ 0 & 0 & 0 & 0 & \frac{1}{2} & \frac{1}{2} \\ 0 & 0 & 0 & 1 & 0 & 0 \end{pmatrix}.$$

Then

$$\mathbf{S} = \begin{pmatrix} 0 & \frac{1}{2} & \frac{1}{2} & 0 & 0 & 0 \\ \frac{1}{6} & \frac{1}{6} & \frac{1}{6} & \frac{1}{6} & \frac{1}{6} & \frac{1}{6} \\ \frac{1}{3} & \frac{1}{3} & 0 & 0 & \frac{1}{3} & 0 \\ 0 & 0 & 0 & 0 & \frac{1}{2} & \frac{1}{2} \\ 0 & 0 & 0 & 1 & 0 & 0 \end{pmatrix}.$$

9.6.3 Problems with S, the Google matrix G.

The rows of $\mathbf{S}$ sum to one, but we have no reason to believe that $\mathbf{S}$ is primitive. So Brin and Page replace $\mathbf{S}$ by

$$\mathbf{G} := \alpha\mathbf{S} + (1-\alpha)\frac{1}{n}\mathbf{J}$$

where $\mathbf{J}$ is the matrix all of whose entries are 1, and $0 < \alpha < 1$ is a real number. (They take $\alpha = 0.85$).

For example, if we start with the 6×6 matrix $\mathbf{H}$ as above, and take $\alpha = .9$, the corresponding Google matrix $\mathbf{G}$ is

$$\begin{pmatrix} 1/60 & 7/15 & 7/15 & 1/60 & 1/60 & 1/60 \\ 1/6 & 1/6 & 1/6 & 1/6 & 1/6 & 1/6 \\ 19/60 & 19/60 & 1/60 & 1/60 & 19/60 & 1/60 \\ 1/60 & 1/60 & 1/60 & 1/60 & 7/15 & 7/15 \\ 1/60 & 1/60 & 1/60 & 7/15 & 1/60 & 7/15 \\ 1/60 & 1/60 & 1/60 & 11/12 & 1/60 & 1/60 \end{pmatrix}.$$

The rows of $\mathbf{G}$ sum to one, and are all positive. So, in principle, $\mathbf{G}^k$ converges to a matrix whose rows are all equal to $\mathbf{s}$ where $\mathbf{s}$ is a solution to

$$\mathbf{s} = \mathbf{s}\cdot\mathbf{G}.$$

MATLAB gives the eigenvalues of G as

$$-0.3705,\ -0.0896,\ 0.6101,\ 1.0000,\ -0.4500,\ -0.4500.$$

The row vector giving the (left) eigenvector with eigenvalue 1 normalized to have row sum 1 is

$$(0.0372,\ 0.0540,\ 0.0415,\ 0.3751,\ 0.2060,\ 0.2862).$$

MATLAB computes G^{10} as

$$\begin{pmatrix} 0.0394 & 0.0578 & 0.0440 & 0.3714 & 0.2044 & 0.2829 \\ 0.0384 & 0.0560 & 0.0429 & 0.3728 & 0.2053 & 0.2846 \\ 0.0389 & 0.0568 & 0.0435 & 0.3707 & 0.2060 & 0.2841 \\ 0.0370 & 0.0535 & 0.0412 & 0.3769 & 0.2049 & 0.2865 \\ 0.0370 & 0.0535 & 0.0412 & 0.3766 & 0.2052 & 0.2865 \\ 0.0370 & 0.0535 & 0.0412 & 0.3732 & 0.2083 & 0.2868 \end{pmatrix}.$$

This is close to, but not quite the limiting value.

MATLAB computes G^{20} as

$$\begin{pmatrix} 0.0372 & 0.0540 & 0.0415 & 0.3751 & 0.2060 & 0.2862 \\ 0.0372 & 0.0540 & 0.0415 & 0.3751 & 0.2060 & 0.2862 \\ 0.0372 & 0.0540 & 0.0415 & 0.3751 & 0.2060 & 0.2862 \\ 0.0372 & 0.0540 & 0.0415 & 0.3751 & 0.2060 & 0.2862 \\ 0.0372 & 0.0540 & 0.0415 & 0.3751 & 0.2060 & 0.2862 \\ 0.0372 & 0.0540 & 0.0415 & 0.3751 & 0.2060 & 0.2862 \end{pmatrix},$$

which has the correct limiting value to four decimal places in all positions. This is what we would expect, since if we take the second largest eigenvalue, which is 0.6101, and raise it to the 20th power, we get .000051.. . In our example, we have seen that the stationary vector with row sum equal to one is

$$\mathbf{s} = \begin{pmatrix} .03721 & .05396 & .04151 & .3751 & .206 & .2862 \end{pmatrix}.$$

The interpretation of the first entry, for example, is that 3.721% of the time, the random surfer visits page 1. The pages of this tiny web are therefore ranked by their importance as (4,6,5,2,3,1).
But, in real life, where 6 is replaced by 12 billion, as **G** is not sparse, taking powers of **G** is impossible due to its size.

9.6.4 Avoiding multiplying by G.

We can avoid multiplying with **G**. Instead, use the iterations scheme

$$\begin{aligned} \mathbf{s}_{k+1} &= \mathbf{s}_k \cdot \mathbf{G} \\ &= \alpha \mathbf{s}_k \cdot \mathbf{S} + \frac{1-\alpha}{n} \mathbf{s}_k \mathbf{J} \\ &= \alpha \mathbf{s}_k \cdot \mathbf{H} + \frac{1}{n}(\alpha \mathbf{s}_k \cdot \mathbf{a} + 1 - \alpha)\mathbf{e} \end{aligned}$$

since $\mathbf{J} = \mathbf{e}^\dagger \otimes \mathbf{e}$ and $\mathbf{s_k} \cdot \mathbf{e}^\dagger = 1$. Now only sparse multiplications are involved.

Why does this converge and what is the rate of convergence?

Let $1, \lambda_2, \ldots$ be the spectrum of **S** and let $1, \mu_2, \ldots$ be the spectrum of **G** (arranged in decreasing order, so that $\lambda_2 < 1$ and $\mu_2 < 1$). We will show that

Theorem 9.6.1.

$$\lambda_i = \alpha\mu_i, \qquad i = 2, 3, \ldots, n.$$

This implies that $\lambda_2 < \alpha$ since $\mu_2 < 1$. Since

$$(0.85)^{50} \doteq 0.000296$$

this shows that at the 50th iteration one can expect 2-3 decimal places of accuracy.

Proof of the theorem. Let $\mathbf{f} := \mathbf{e}^\dagger$ so $\mathbf{f}$ is the column vector all of whose entries are 1. Since the row sums of $\mathbf{S}$ equal 1, we have $\mathbf{S} \cdot \mathbf{f} = \mathbf{f}$. Let $\mathbf{Q}$ be an invertible matrix whose first column is $\mathbf{f}$, so $\mathbf{Q} = (\mathbf{f}, \mathbf{X})$ for some matrix $\mathbf{X}$ with n rows and $n-1$ columns. Write $\mathbf{Q}^{-1}$ as $\mathbf{Q}^{-1} = \begin{pmatrix} \mathbf{y} \\ \mathbf{Y} \end{pmatrix}$ where $\mathbf{y}$ is a row vector with n entries and $\mathbf{Y}$ is a matrix with $n-1$ rows and n columns. The fact that $\mathbf{Q}^{-1}\mathbf{Q} = \mathbf{I}$ implies that

$$\mathbf{y} \cdot \mathbf{f} = 1 \quad \text{and } \mathbf{Y} \cdot \mathbf{f} = \mathbf{0}.$$

We have

$$\mathbf{Q}^{-1}\mathbf{S}\mathbf{Q} = \begin{pmatrix} \mathbf{y} \cdot \mathbf{f} & \mathbf{y}\mathbf{S}\mathbf{X} \\ \mathbf{Y} \cdot \mathbf{f} & \mathbf{Y}\mathbf{S}\mathbf{X} \end{pmatrix} = \begin{pmatrix} 1 & \mathbf{y}\mathbf{S}\mathbf{X} \\ \mathbf{0} & \mathbf{Y}\mathbf{S}\mathbf{X} \end{pmatrix},$$

So the eigenvalues of $\mathbf{YSX}$ are $\lambda_2, \lambda_3, \ldots$. Now $\mathbf{J}$ is a matrix all of whose columns equal $\mathbf{f}$. So $\mathbf{Q}^{-1}\mathbf{J}$ has ones in the top row and zeros elsewhere. So

$$\mathbf{Q}^{-1}\mathbf{J}\mathbf{Q} = \begin{pmatrix} 1 & \mathbf{e} \cdot \mathbf{X} \\ 0 & \mathbf{0} \end{pmatrix}$$

Hence

$$\mathbf{Q}^{-1}\mathbf{H}\mathbf{Q} = \mathbf{Q}^{-1}(\alpha\mathbf{S} + (1-\alpha)\mathbf{J})\mathbf{Q} = \begin{pmatrix} 1 & \alpha\mathbf{y}\mathbf{S}\mathbf{X} + (1-\alpha)\mathbf{e} \cdot \mathbf{X} \\ 0 & \alpha\mathbf{Y}\mathbf{S}\mathbf{X} \end{pmatrix}.$$

So the eigenvalues of $\mathbf{G}$ are $1, \alpha\lambda_2, \alpha\lambda_3 \ldots$. □

9.7 Eigenvalue sensitivity and reproductive value.

Let A be a primitive matrix, r its maximal eigenvalue, x a right eigenvector with eigenvalue r, y a left eigenvector with eigenvalue r with $y \cdot x = 1$ and H the one dimensional projection operator $H = x \otimes y$ so

$$H = \lim_{k \to \infty} \left(\frac{1}{r} A\right)^k.$$

If e_j is the (column) vector with 1 in the j-th position and zeros elsewhere, then

$$He_j = y_j x.$$

This equation has a "biological" interpretation due to R.A.Fisher: If we think of the components of a column vector as referring to stages of development (as, for example, in the Leslie matrix), then the components of y can be interpreted as giving the relative "reproductive value" of each stage:

Think of different stages as alternate investment opportunities in long-term population growth. If you could put one dollar into any one of these investments (one individual in any of the stages) what is their relative payoff in the long run (the relative size of the resulting population in the distant future)? The above equation shows that it is proportional to y_j.

Eigenvalue sensitivity to changes in the matrix elements.

The Perron-Frobenius theorem tells us that if we increase any matrix element in a primitive matrix, A, then the dominant eigenvalue r increases. But by how much? To answer this question, consider the equation

$$y \cdot A \cdot x = r\ y \cdot x = r.$$

In this equation, think of the entries of A as n^2 independent variables, and x, y, r as functions of these variables.

Take the partial derivative with respect to the ij-th entry, a_{ij}. The left hand side gives

$$\frac{\partial y}{\partial a_{ij}} \cdot A \cdot x + y \cdot \frac{\partial A}{\partial a_{ij}} \cdot x + y \cdot A \cdot \frac{\partial x}{\partial a_{ij}}.$$

But $\frac{\partial A}{\partial a_{ij}}$ is the matrix with 1 in the ij-th position and zeros elsewhere, and the sum of the first and third terms above are (since $Ax = rx$ and $yA = ry$)

$$r\left(\frac{\partial y}{\partial a_{ij}} \cdot x + y \cdot \frac{\partial x}{\partial a_{ij}}\right) = r\frac{\partial (y \cdot x)}{\partial a_{ij}} = 0$$

since $y \cdot x \equiv 1$. So we have proved that

$$\frac{\partial r}{\partial a_{ij}} = y_i x_j. \tag{9.4}$$

I will now present Fischer's use of this equation to "explain" why we age. The following discussion is taken almost verbatim from the book [Ellner and Guckenheimer] pages 50-51. This explanation is derived by modeling a life cycle in which there is no aging, and then asking whether a little bit of aging would lead to increased Darwinian fitness as measured by r.

Chapter 10

Some topics in ordinary differential equations.

This is not a text on ordinary differential equations. By contrast, a course on dynamical systems given 40 years ago would consist almost entirely in the study of ordinary differential equations. There are many excellent ode texts. In this chapter I cover those topics in ode which I regularly teach in my dynamical systems course. In the next chapter I deal with a more specialized topic - the Lotka-Volterra equations and their generalizations.

We have proved the local existence theorem for equations of the form

$$x'(t) = F(t,x), \quad x(0) = x_0$$

under Lipschitz assumptions on F via the contraction fix point theorem. I begin with an important special case.

10.1 Linear equations with constant coefficients.

10.1.1 Linear homogenous equations with constant coefficients.

These are equations of the form

$$x' = Ax$$

where A is a constant bounded linear operator on a Banach space, X. You may as well think of X as a finite dimensional vector space. We must also specify the initial conditions, of course.

In case $X = \mathbb{R}$, so $A = a$ is a scalar, we know that the general solution to the above equation is $x(t) = ce^{at}$ where the constant c is determined by the initial condition, $c = x(0)$ and

$$e^{at} = 1 + at + \frac{1}{2}a^2t^2 + \frac{1}{3!}a^3t^3 + \cdots,$$

where this series converges for all t.

Exactly the same method works in general!

The exponential series for an operator.

Define

$$e^{tA} = I + tA + \frac{1}{2}t^2A^2 + \frac{1}{3!}t^3A^3 + \cdots.$$

Here I is the identity operator so $e^{0A} = I$ is the identity. The exact same proof of the convergence of the exponential series in one variable shows that this series converges for all t and that

$$e^{(s+t)A} = e^{sA}e^{tA}.$$

It is *not true* in general that $e^{tA}e^{tB} = e^{t(A+B)}$. This lack of equality stems from the fact that A and B may not commute, and this fact is manifested in the physical world by quantum mechanics. But if A and B do commute then $e^{tA}e^{tB} = e^{t(A+B)}$.

The derivative of the exponential.

We may differentiate the exponential series with respect to t term by term and we find that

$$\frac{d}{dt}e^{tA} = Ae^{tA} = e^{tA}A.$$

In particular, if we set

$$x(t) = e^{tA}x_0$$

then

$$x'(t) = Ax(t) \quad \text{and} \quad x(0) = x_0.$$

So the study of linear homogeneous differential equations with constant coefficients reduces to the analysis of e^{tA}.

$e^{tPAP^{-1}} = Pe^{tA}P^{-1}$.

Suppose that P is an invertible operator and

$$B = PAP^{-1}.$$

then $B^2 = PA^2P^{-1}$, $B^3 = PA^3P^{-1}$ etc. so

$$e^{tB} = Pe^{tA}P^{-1}.$$

We know from linear algebra that (in finite dimensions) every B is of the form $B = PAP^{-1}$ where A has a "nice" normal form. So our study is reduced to understanding e^{tA} for normal forms (and also understanding the effect of conjugating by P).

Here is a complete analysis in two (real) dimensions:

10.1.2 e^{tB} where B is a two by two real matrix.

1. B has two real distinct eigenvalues.

Then B can be diagonalized, i.e. $B = PAP^{-1}$ where

$$A = \begin{pmatrix} a & 0 \\ 0 & b \end{pmatrix}$$

and

$$e^{tA} = \begin{pmatrix} e^{ta} & 0 \\ 0 & e^{tb} \end{pmatrix}.$$

The x and y axes are invariant.

Further analysis depends on signs of a and b:

- If $a < 0$ and $b < 0$ then e^{tA} for $t > 0$ contracts all points (at different rates) toward the origin. The effect of the conjugation by P is to replace the x and y axes by other lines through the origin.
- If $a > 0$ and $b < 0$ then e^{tA} is hyperbolic for $t \neq 0$. For $t > 0$ e^{tA} is expanding in the x-direction and contracting in the y-direction (with the reverse for $t < 0$). Again, the effect of the conjugation by P is to replace the x and y axes by other lines through the origin.
- If $a = 0$ and $b < 0$ then points on the x-axis are stationary and points all converge toward the x-axis as $t \to +\infty$.
- Of course, if $a = b = 0$ so that $A = 0$, $e^{tA} \equiv I$.

This covers all cases (up to interchange of a and b and t and $-t$.)

B has two equal real eigenvalues.

Here there are two cases:

- If
  $$A = \begin{pmatrix} a & 0 \\ 0 & a \end{pmatrix}$$
  then
  $$e^{tA} = e^{ta}I.$$
  So uniform expansion (for $t > 0$) if $a > 0$, uniform contraction if $a < 0$ and $e^{tA} \equiv I$ if $a = 0$. Since an A of this form commutes with all other matrices, if $B = PAP^{-1}$ then $B = A$.
- The second case is
  $$A = \begin{pmatrix} a & 1 \\ 0 & a \end{pmatrix}.$$

This is of the form $A = aI + C$ where

$$C = \begin{pmatrix} 0 & 1 \\ 0 & 0 \end{pmatrix}$$

so

$$e^{tA} = e^{ta}e^{tC}.$$

To compute e^{tC} observe that $C^2 = 0$ so

$$e^{tC} = \begin{pmatrix} 1 & t \\ 0 & 1 \end{pmatrix}.$$

Non-real eigenvalues.

If the eigenvalues are $a \pm ib$ with $b \neq 0$ then a normal form is

$$A = \begin{pmatrix} a & b \\ -b & 0 \end{pmatrix} = aI + b\begin{pmatrix} 0 & 1 \\ -1 & 0 \end{pmatrix}.$$

So if

$$C = \begin{pmatrix} 0 & 1 \\ -1 & 0 \end{pmatrix}$$

then

$$e^{tA} = e^{ta}e^{tbC}$$

so we must compute e^{tC}.

We claim that

$$e^{t\begin{pmatrix} 0 & 1 \\ -1 & 0 \end{pmatrix}} = \begin{pmatrix} \cos t & \sin t \\ -\sin t & \cos t \end{pmatrix}.$$

This follows from

$$C^2 = -I \quad \text{so } C^3 = -C, \;\; C^4 = I.$$

Thus

$$e^{tC} = \begin{pmatrix} 1 - \frac{1}{2}t^2 + \frac{1}{4!}t^4 + \cdots & t - \frac{1}{3!}t^3 + \cdots \\ & \\ -t + \frac{1}{3!}t^3 - \cdots & 1 - \frac{1}{2}t^2 + \frac{1}{4!}t^4 + \cdots \end{pmatrix}.$$

Thus if $a = 0$, the trajectories of e^{tA} are circles with velocity of rotation b. If $a < 0$ the trajectories are circular spirals heading into the origin as $t \to \infty$, and if $a > 0$ the trajectories are circular spirals spiraling out.

The effect of conjugating by P is to replace the circles by ellipses and circular spirals by elliptical spirals.

10.2 Hyperbolicity for differential equations.

Let A be an $n \times n$ real matrix. If no eigenvalue of A has real part zero, then for any $t \neq 0$, the linear map e^{tA} is hyperbolic in the sense of Chapter 8.

If

$$\frac{dx}{dt} = F(x)$$

is a system of ordinary differential equations on $\mathbb{R}^n$ with $F(0) = 0$ and $F'(0) = A$, my old conjugacy theorems say that near 0 the solutions of this system are conjugate to the corresponding linear system: Conjugacy via a homeomorphism in general, and via smooth maps if appropriate non-resonance conditions are satisfied. In fact, the differential equations case can be reduced to the discrete mapping case. I will not go into this subject further here. For a nice discussion of these theorems from the point of view of quantum mechanical scattering theory, see [Nelson].

10.3 Bifurcations of differential equations.

Just as we studied in the behavior of maps in one dimension near fixed points with $f'(p) = \pm 1$ there is a similar study of bifurcations of non-hyperbolic zeros of vector fields. A famous example is the *Hopf bifurcation.* Here is an illustration of this phenomenon, without going into the technical details. Suppose our system of differential equations is

$$\begin{aligned} x' &= ax + y - x(x^2 + y^2) \\ y' &= ay - x - y(x^2 + y^2). \end{aligned}$$

If $a < 0$, all trajectories spiral into the origin. If $a > 0$ and small, then the origin has become an unstable fixed point, and all trajectories starting near the origin spiral outward towards a periodic trajectory while points outside the periodic trajectory spiral in towards it.

In other words, as a passes from negative to positive, an attractive fixed point has become repulsive and a nearby attractive periodic orbit has appeared.

The study of bifurcations of differential equations is another extremely important topic which we will omit. Once again there are many excellent texts, for example [Kuznetsov].

10.4 Variation of constants.

We studied the homogeneous equation

$$x'(t) = Ax(t), \quad x(0) = x_0.$$

Suppose we want to solve the "inhomogeneous" equation

$$\frac{d}{dt}x(t) = Hx(t) + f(t)$$

where f is given, and with the initial condition $x(0) = x_0$. The solution is given by Lagrange's **variation of constants formula**

$$x(t) = e^{tH}x_0 + \int_0^t e^{(t-s)H} f(s)ds$$

as can be checked by differentiating the right hand side. This formula is also known as **Duhamel's formula.**

10.4.1 The operator version.

If F is an operator valued function of t then the operator version of the above says that

$$X(t) = e^{tH} + \int_0^t e^{(t-s)H} F(s)ds \tag{10.1}$$

is the solution to the differential equation

$$\frac{d}{dt}X = HX + F$$

with the initial conditions

$$X(0) = I.$$

The Born series.

For example, suppose that $H = H_0 + H_1$ and we take $F(t) = H_1 e^{tH}$. We want to find a solution to

$$\frac{d}{dt}Y(t) = HY(t) = H_0Y(t) + H_1Y(t), \qquad Y(0) = I.$$

The variation of constants formula tell us that

$$Y(t) = e^{tH_0} + \int_0^t e^{(t-s)H_0} H_1 e^{sH} ds. \tag{10.2}$$

If we substitute this formula into itself (i.e. use this formula for the e^{sH} occurring in the integral on the right) we get

$$e^{tH} = e^{tH_0} + \int_0^t e^{(t-s)H_0} H_1 e^{sH_0} dt + \int_0^t \int_0^s e^{(t-s)H_0} H_1 e^{(s-\tau)H_0} H_1 e^{\tau H} d\tau ds.$$

Clearly we can keep going. The usefulness of this scheme is as follows. Suppose that H_1 is small, for example suppose that we can ignore all terms involving three products of H_1. Then in the above expression, replacing $e^{\tau H}$ by $e^{\tau H_0}$ on the right, we get an approximate expression for e^{tH} in terms of integrals involving e^{tH_0} and products of H_1. The corresponding series (and approximation) is known as the Volterra series (or approximations) to mathematicians, and is known as the Born series (or approximations) to physcists.

10.4.2 The parametrix expansion.

Suppose we want to find e^{tH} and we only found an approximate solution - we have found an operator valued function $K(t)$ such that

$$\frac{dK(t)}{dt} = HK(t) + R(t), \qquad K(0) = I.$$

The variation of constants formula tells us that

$$K(t) = e^{tH} + \int_0^t e^{(t-s)H} R(s)ds$$

which we shall write as

$$e^{tH} = K(t) - \int_0^t e^{(t-s)H} R(s)ds.$$

Substitute this back into itself to obtain

$$e^{tH} = K(t) - \int_0^t K(t-s)R(s)ds + \int_0^t \int_0^{t-s} e^{(t-s-\tau)H} R(\tau)R(s)d\tau ds.$$

Keep going. This suggests the following: LetΔ_k denote the k-simplex

$$\Delta_k = \{(t_1, \ldots, t_k | 0 \le t_1 \le t_2 \le \cdots \le t_k \le 1\}.$$

If all the t_i are unequal, there are $k!$ ways of of reordering them. Since we may ignore possible equalities in computing volume, this shows that the Euclidean volume ofΔ_k is $1/k!$.
So the volume of $t\Delta_k$ is $t^k/k!$. Define the operators $Q(k,t)$ by

$$Q(k,t) := \int_{t\Delta_k} K(t-t_k)R(t_k - t_{k-1}) \cdots R(t)dt_1 \cdots dt_k.$$

So this integral is over $0 \le t_1 \le t_2 \cdots \le t_k \le t$. To shorten the formulas, I will drop the $dt_1 \cdots dt_k$.

If K and R are uniformly bounded, say by C, in the interval $[0,T]$, then the $Q(k,t)$ are bounded by $C^k t^k/k!$ and so the series

$$\sum_{k=0}^{\infty} (-1)^k Q(k,t)$$

converges uniformly. $R(k,s)$ by

$$R(k,s) = \int_{s\Delta_{k-1}} R(s - t_{k-1})rR(t_{k-1} - t_{k-2}) \cdots R(t_2 - t_1)R(t_1)$$

so that

$$Q(k,t) = \int_0^t K(t-s)R(k,s)ds.$$

If we apply $\left(\frac{d}{dt} - H\right)$ to $Q(k,t)$ we get

$$R(k,s) + R(k+1,s)$$

so the sum in

$$\left[\frac{d}{dt} - H\right]\left(\sum_{k=0}^{\infty}(-1)^k Q(k,t)\right)$$

telescopes to zero. Hence

$$\sum_{k=0}^{\infty}(-1)^k Q(k,t)$$

is a solution to our search for e^{tH} starting with an approximate solution. This method is known as the **parametrix expansion**.

The rest of this chapter is devoted to the study of autonomous differential equations, that is differential equations of the form

$$\frac{dx}{dt} = F(x).$$

It is usual to consider the map $x \mapsto F(x)$ as a vector field, i.e a rule which assigns a vector to each point x, and to emphasize this viewpoint, we will frequently write V instead of F.

10.5 The Poincaré-Bendixon theorem.

The global behavior of bounded trajectories of an autonomous system of differential equations in the plane (i.e. of a time independent vector field, V) have a deceptively simple beautiful structure given by the Poincaré Bendixon theorem which we shall state after a definition:

10.5.1 The ω-limit set.

If $C = C(t)$ is a trajectory of a time independent vector field, we define the **omega limit set** of C, denoted by $\omega(C)$, to consist of all points p such that there exists a sequence $t_n \to \infty$ such that

$$C(t_n) \to p.$$

We will also use the notation $\omega(x)$ for $\omega(C)$ when $x = C(t)$ for some t. These definitions are valid in all dimensions.

10.5.2 Statement of the Poincaré-Bendixon theorem.

Theorem 10.5.1. *If C is a (forward) bounded trajectory of a vector field, V, in the plane, and if $\omega(C)$ contains no zeros of V then either:*
(1) C $(=\omega(C))$ is a periodic trajectory, or

(2) $\omega(C)$ consists of a periodic trajectory of V which C approaches spirally from the inside or from the outside.

Suppose that V has a finite number of zeros and that $\omega(C)$ contains a zero, A, of V. Then

(3) If $\omega(C)$ consists only of zeros of V, then, in fact, $\omega(C)$ consists of the single point, A, and $C(t)$ approaches A as $t \to \infty$. Otherwise

(4) $\omega(C)$ consists of a finite set $\{A_n\}$ of zeros of V and a set of trajectories, $\{C_a\}$, where, as $t \to \pm\infty$ each C_a approaches one of the zeros.

At a crucial point in the proof of the theorem we will need to make use of the Jordan curve theorem in the plane, and this is why the theorem is peculiar to the plane. But many of the preliminary results are of independent interest and are valid in any number of dimensions. So I will begin with these more general results, and will warn you when we come to special properties of the plane.

10.5.3 Properties of the omega limit set of a trajectory, in the general case.

Lemma 10.5.1. *The set $\omega(C)$ is closed and invariant under the flow generated by V.*

Proof that $\omega(C)$ is closed. Suppose that $\{A_n\}$ is a sequence of points in $\omega(C)$ which converge to a point A. We can find a sequence $\{t_n\}$ such that $t_n > n$ and $d(C(t_n), A_n) < 1/n$ where d denotes distance. Then $C(t_n) \to A$, proving that $\omega(C)$ is closed.

Proof that $\omega(C)$ is invariant. Suppose that $A \in \omega(C)$ and let $D(t)$ be the trajectory through A with $D(0) = A$. Let $\{t_n\} \to \infty$ be a sequence with $C(t_n) \to A$. Define

$$C_n(t) = C(t + t_n)$$

so that C_n is a reparametrization of the trajectory, C with

$$C_n(0) \to A.$$

For any fixed t, the continuous dependence of solutions of differential equations upon initial conditions implies that

$$C_n(t) \to D(t).$$

But this is the same as saying that $C(t_n + t) \to D(t)$ proving that $D(t) \in \omega(C)$. □

Lemma 10.5.2. *If the forward trajectory of C is bounded, then $\omega(C)$ is connected.*

Proof. By the preceding lemma, $\omega(C)$ is closed, and by hypothesis it is bounded. If it is not connected, then we can represent it as the union of two disjoint closed subsets, $\omega(C) = M \cup N$. Since M and N are closed, bounded, and disjoint, they are at a positive distance from one another. Call this distance δ. Now we can choose a monotonic sequence $t_n \to \infty$ such that for all n sufficiently large,

$$d(C(t_n), M) < \delta/4, \quad n \text{ odd}, \quad d(C(t_n), N) < \delta/4, \quad n \text{ even}.$$

By continuity, we can thus find a sequence $s_k \to \infty$ with

$$t_{2k-1} < s_k < t_{2k} \quad \text{and} \quad d(C(s_k), M \cup N) \geq \delta/2.$$

But this leads to a contradiction, since the sequence of points $C(s_k)$ is bounded, and hence must contain a convergent subsequence whose limit, A, belongs to $\omega(C)$ by definition, and yet is a postive distance from $\omega(C) = M \cup N$. □

Notice that this lemma implies part 3) of the Poincaré Bendixon theorem. Indeed, since V is assumed to have only finitely many zeros, if $\omega(C)$ consists only of zeros, it must consist of exactly one point if it is to be connected.

So this assertion is valid an any dimension.

Transversal surfaces (curves) to a vector field.

By a **transversal**, L, to the vector field V we mean a surface of codimension one which is nowhere tangent to V. In the plane, this means that L is a curve.

In particular, the vector field, V does not vanish at any point of L. If $V(A) \neq 0$, we can always find a transversal to V passing through A: Simply choose a subspace of codimension one of the tangent space at A which does not contain $V(A)$, and then choose a surface tangent to this subspace at A. At all points sufficiently near to A the vector field V will not be tangent to this surface on account of continuity.

We will only be dealing with transversals locally, and locally every surface has two sides. The vector field V points towards one of these two sides at A, and hence by continuity must point to this same side at all points of the transversal. In other words, the trajectories all cross the transversal in the same direction.

Lemma 10.5.3. *For any bounded interval $[a, b]$ a trajectory C can cross the transversal, L, at most a finite number of times for $a \leq t \leq b$.*

Proof. Suppose the contrary. We would then have an infinite sequence of times $a \leq t_n \leq b$ with $C(t_n) \in L$. By passing to a subsequence, we may assume that the t_n converge to some point $s \in [a, b]$. So the points $C(t_n)$ lie on L and converge to $C(s) \in L$. If more than one of the points $C(t_n)$ coincides with $C(s)$, then C is a periodic trajectory, of some minimal period, say h. It can not be the case that infinitely many of the points t_n have $C(t_n) = C(s)$, because the difference between each successive such t_n has to be a multiple of h, and so these infinitely many t_n's could not lie in the bounded interval $[a, b]$. So in all cases, we will have infinitely many of the t_n with $C(t_n) \neq C(s)$. Passing to this subsequence if necessary, we find a collection of secants (in local coordinates) $\overline{C(t_n)C(s)}$ whose limiting direction is tangent to C, but is also tangent to L, a contradiction. □

Lemma 10.5.4. *Let A be a point of a transversal, L. For every $\epsilon > 0$ there is a neighborhood U of A such that every trajectory, C with $C(0) \in U$ intersects L at some time t with $|t| < \epsilon$.*

By choice of local coordinates we can arrange

- That $A = 0$,
- that the transversal is given by the equation $y = 0$ in terms of local coordinates $(x^1, \dots, x^{n-1}, y) = (x, y)$,
- and that the y coordinate of the vector field, V is positive.

With these choices we now proceed to the proof of the lemma:

Proof. Let $C(x_0, y_0, t)$ denote the trajectory which passes through (x_0, y_0) at $t = 0$, and let $y(x_0, y_0, t)$ denote the y coordinate of $C(x_0, y_0, t)$. Transversality says that

$$\frac{\partial y}{\partial t}(0, 0, 0) > 0.$$

By the implicit function theorem, the equation

$$y(x_0, y_0, t) = 0$$

has a unique continuous solution $t(x_0, y_0)$ with $|t| < \epsilon$ for (x_0, y_0) in some neighborhood of the origin. □

We now come to results which are particular to the plane.

10.6 Proof of Poincaré-Bendixon.

So from now on V is vector field in the plane, and L will be a closed line segment which is part of an open segment which is a transversal to V. We shall call L a trnasversal segment.

Lemma 10.6.1. *Suppose $\omega(C)$ contains a point A with $V(A) \neq 0$. If L is a transversal segment through A, there exists a monotone sequence of times $t_n \to \infty$ such that the points of intersection of $C(t)$ with L for $t \geq 0$ are precisely the points $C(t_n)$. If $C(t_1) = C(t_2)$ then $C(t_n) = A$ for all n and C is*

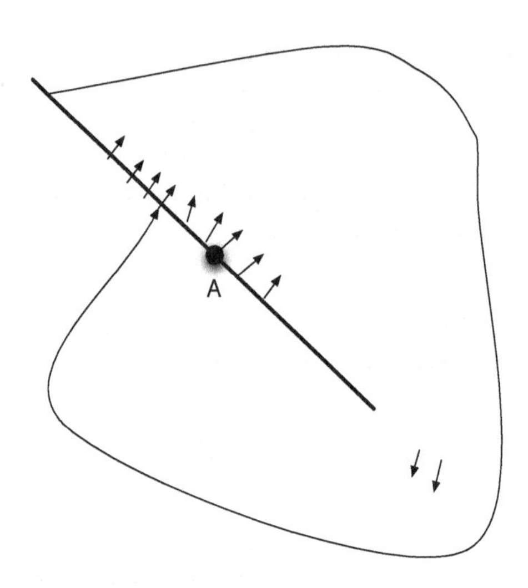

Figure 10.1: Crossing a transversal in the plane.

periodic. If $C(t_1) \neq C(t_2)$ then all the points $C(t_n)$ are distinct, $C(t_{n+1})$ lies between $C(t_n)$ and $C(t_{n+2})$ on L and the sequence of points $C(t_n)$ converges monotonically on L to A. In particular, $\omega(C) \cap L = \{A\}$. In other words, the transversal L contains only the single point A of $\omega(C)$.

Proof. **Step 1.** By definition, every neighborhood of A contains points of the form $C(t)$ with arbitrarily large t. Hence by the preceding lemma, the curve $C(t)$ will cross L infinitely many times with arbitrarily large t. By the lemma before last, any finite interval of time contains only finitely many such intersections, and so the intersection times are given by a monotone increasing sequence as stated in the lemma. If $C(t_1) = C(t_2)$ then C is periodic with period $t_2 - t_1$, and, by definition the curve C does not cross L at any time between t_1 and t_2. So $C(t_n) = C(t_1)$ and as A is the limit of the $C(t_n)$ (by the preceeding lemma) we conclude that $A = C(t_1)$. (So far we have not used properties particular to the plane.)

Step 2. Suppose that $C(t_1) \neq C(t_2)$. By definiton, C does not intersect L for $t_1 < t < t_2$. So the curve formed by $C(t)$, $t_1 \leq t \leq t_2$ and the segment $\overline{C(t_1)C(t_2)}$ of L forms a simple closed curve,Γ.

We claim that the trajectory $C(t), t > t_2$ can not crossΓ . Indeed, suppose that for $t > t_2$ but close to t_2 the curve $C(t)$ lies insideΓ . It can not cross the C portion of Γ by the uniqueness theorem of ordinary differential equations, and it can not cross the L portion in the direction opposite to the trajectories at t_1 and t_2.

Hence it lies entirely inside Γ for all time. In particular, $C(t_3)$ is insideΓ and $C(t_2)$ lies between $C(t_1)$ and $C(t_3)$ on L. By induction, the $C(t_n)$ form

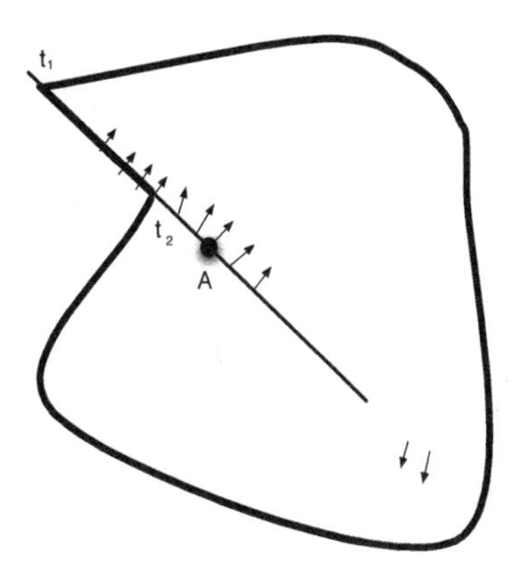

Figure 10.2: The curveΓ.

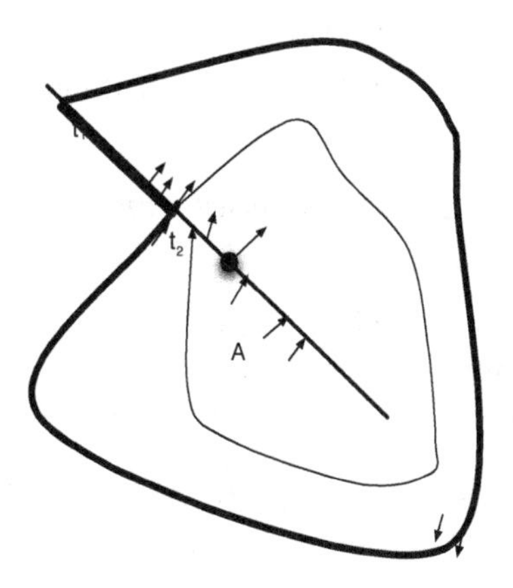

Figure 10.3: C is trapped insideΓ.

a monotone sequence on L. If the curve starts outside Γ the same argument shows that it must remain outside for all time and the same monotonicity holds, completing the proof of the lemma.

□

We now turn to the proof of the Poincaré-Bendixon theorem and proceed according to cases.

Suppose that $\omega(C)$ contains non zeros of V.

Let A be a point of $\omega(C)$, let D be the trajectory through A and consider $\omega(D)$. Since $\omega(C)$ is closed and since $D(t) \in \omega(C)$ for large t, we conclude that $\omega(D) \subset \omega(C)$. If we suppose that $\omega(C)$ contains no zeros of V, then a point $B \in \omega(D)$ is not a zero of V. The set $\omega(D)$ is not empty since the entire forward trajectory through A is bounded, being contained in $\omega(C)$. Choose a transverse segment L through B. By the preceeding lemma, L can intersect $\omega(C)$ in only one point. In particular, L can intersect D in at most one point, and hence we conclude that D is periodic. If C is periodic, we have $C = D$ and we are in case 1) of the Poincaré Bendixon theorem.

If C is not periodic, it must lie either entirely inside or entirely outside the closed curve, F given by the trajectory of D. Choose a point $A \in F$ and a transverse segment thrrough A. Then the preceeding lemma (together with the Jordan curve theorem) shows that C spirals towards F giving case 2) of the theorem.

Notice that our argument showed that if D is a trajectory contained in $\omega(C)$ and if $\omega(D)$ contains a point, B with $V(B) \neq 0$ (so that we can pass a transversal through B) then D must be periodic.

We have already proved part 3) in arbitrary dimensions. For the last assertion we need

Lemma 10.6.2. *If $\omega(C)$ contains a closed trajectory, D, it contains no other points.*

Proof. Suppose the contrary. Then $\omega(C)/D$ is not empty, and it can not be closed since $\omega(C)$ is connected. So there is a point A lying on D which is a limit point of points of $\omega(C)/D$. Let L be a transverse segment through A. Since every neighborhood of A contains points of $\omega(C)/D$, it follows that the trajectories of these points close enough to A cross L. But all of the points on these trajectories lie in $\omega(C)$. Hence L contains more than one point of $\omega(C)$, a contradiction. □

Completion of the proof of Poincaré-Bendixon.

Now suppose that $\omega(C)$ contains a finite number of zeros of V and also some points where $V(A) \neq 0$. Consider the trajectory through the point A. It can not be periodic by the preceding lemma. But then, by the above argument, its omega limit set can not contain a point where V does not vanish. Hence by part 3), its omega limit set consists of a single zero, which must be one of the finitely many zeros contained in $\omega(C)$. Reversing time, we see that as $t \to -\infty$ the trajectory also tends to a zero in our set. $\square$

10.7 The van der Pol and Lienard equations.

10.7.1 The van der Pol equation.

Van der Pol introduced his equation in 1920 to describe oscillations in a triode circuit. He and van der Mark used this equation to describe the heart beat in a paper entitled "The heartbeat considered as a relaxation oscillation, and an electrical model of the heart" which appeared in *The London, Edinburgh, and Dublin Philosophical Magazine and Journal of Science Ser.7,* **6** (1928) 763-775. The equation is

$$\ddot{x} + x = \mu(1 - x^2)\dot{x} \tag{10.3}$$

where $\mu > 0$ is a positive parameter. Bringing the right hand side over to the left we can write this as

$$\ddot{x} + f(x)\dot{x} + x = 0, \tag{10.4}$$

where $f(x) := \mu(x^2 - 1)$. The properties of f that we shall use are:

a. $F(x) = \int_0^x f(s)ds$ is an odd function. In particular, $F(0) = 0$.

b. $F(x) \to \infty$ as $x \to \infty$ and there is a zero , z, of F so that for $x > z$, $F(x) > 0$ and F is monotone increasing.

c. For $0 < x < z$, we have $F(x) < 0$.

(For the case of the van der Pol equation, $z = \sqrt{3}$.)

10.7.2 The Lienard equations.

Let us set $y = \dot{x} + F(x)$ so the equation (10.4) becomes the system

$$\dot{x} = y - F(x) \tag{10.5}$$

$$\dot{y} = -x \tag{10.6}$$

These equations are known as the Lienard equations, where properties a-c are assumed. They imply that origin is the only critical point, i.e. the only zero of the vector field. Indeed, if $\dot{y} = 0$, equation (10.6) says that $x = 0$, and since $F(0) = 0$, equation (10.5) then says that $y = 0$.

Figure 10.4 sketches the vector field corresponding to the van der Pol equation with $\mu = 1$ near the origin. Notice that the vector field seems to generate a flow spiraling outward from the origin and spiraling inward outside a neighborhood of the origin. So we suspect the existence of a limit periodic solution. This can be experimentally verified using Matlab's ode45 differential equation program. See Figure 4.5.

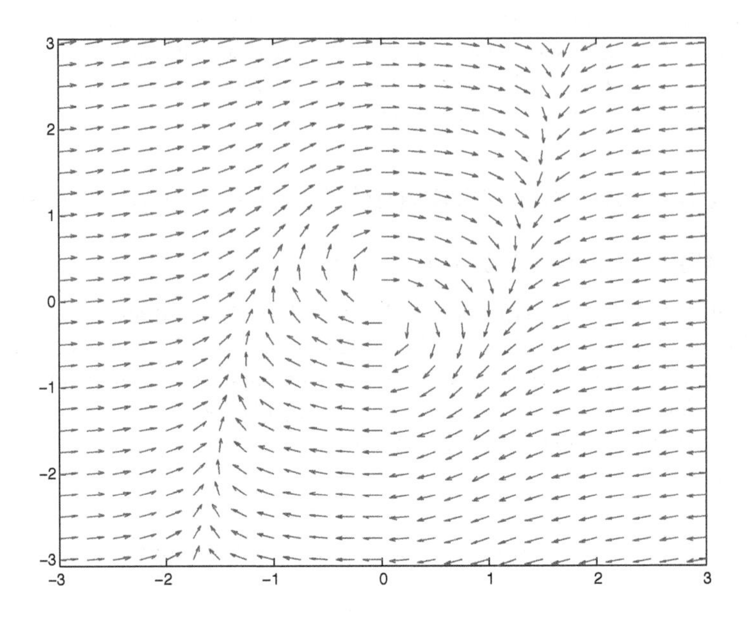

Figure 10.4: The vector field for van der Pol ($\mu = 1$) near the origin.

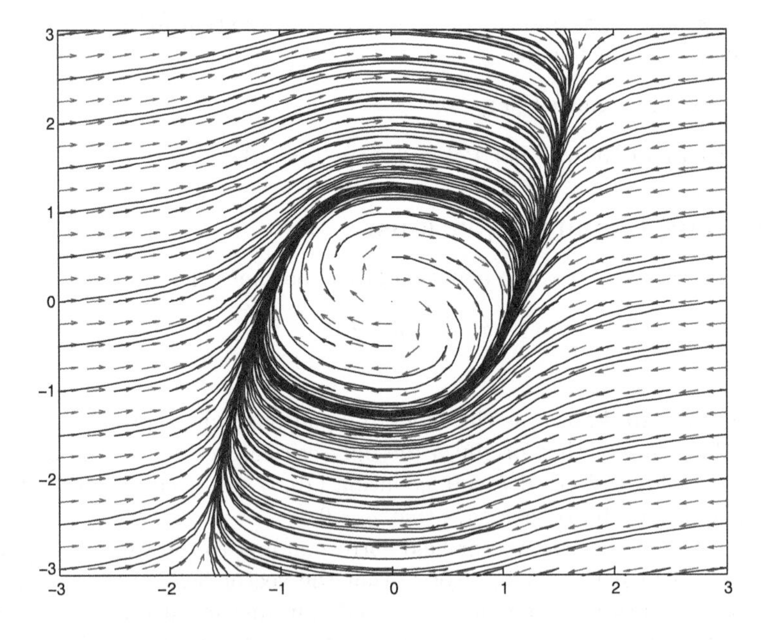

Figure 10.5: Solutions using MATLAB's ode45.

10.7.3 Proofs.

Let us now justify these figures:

Solutions spiral outward from the origin.

This is a purely linear algebra problem. If we show that the matrix A giving the derivative of the vector field at the origin has non-real eigenvalues with positive real part, then, since the behavior near the origin, which is homeomorphic to the solution of the linearized equations, will consist of outward spirals.

At the origin, the matrix of the linearized equation is

$$\begin{pmatrix} -f(0) & 1 \\ -1 & 0 \end{pmatrix}$$

which has trace $-f(0)$ and determinant 1 and so the eigenvalues are

$$-\frac{1}{2}f(0) \pm \frac{1}{2}\sqrt{f^2(0) - 4}.$$

If $f(0) \neq 0$, then conditions a) and c) implies that $f(0) < 0$ and hence both eigenvalues have positive real parts: orbits move out from the origin with increasing time.

A better argument. In fact, we can conclude this from c without any assumptions about $f(0)$: Set

$$R = \frac{1}{2}(x^2 + y^2).$$

Then

$$\begin{aligned} \dot{R} &= x\dot{x} + y\dot{y} \\ &= x(y - F(x)) + y(-x) \\ &= -xF(x) \end{aligned}$$

which is positive for $0 < |x| < z$. So points near the origin increase their radial distance until at least $|x| \geq z$.

Radial decrease from $|x| > z$. The same argument shows that the radial distance is decreasing if $|x| > z$.

Trying to use Poincaré-Bendixon. So if a trajectory is to escape offto infinity, it must do so by passing through the strip $-z \leq x \leq z$. We shall show that this is impossible, and hence conclude that all trajectories remain bounded. According to Poincaré -Bendixon, as there are no attractive zeros of our vector field, the only possibility is a limit cycle. So by proving that all solution remain bounded, we will conclude the existence of a periodic solution. Showing that no trajectory can escape to infinity by passing through the strip $-z \leq x \leq z$ is the hard part of the argument. Our proof will follow the discussion in [Robinson] pp.173 - 176.

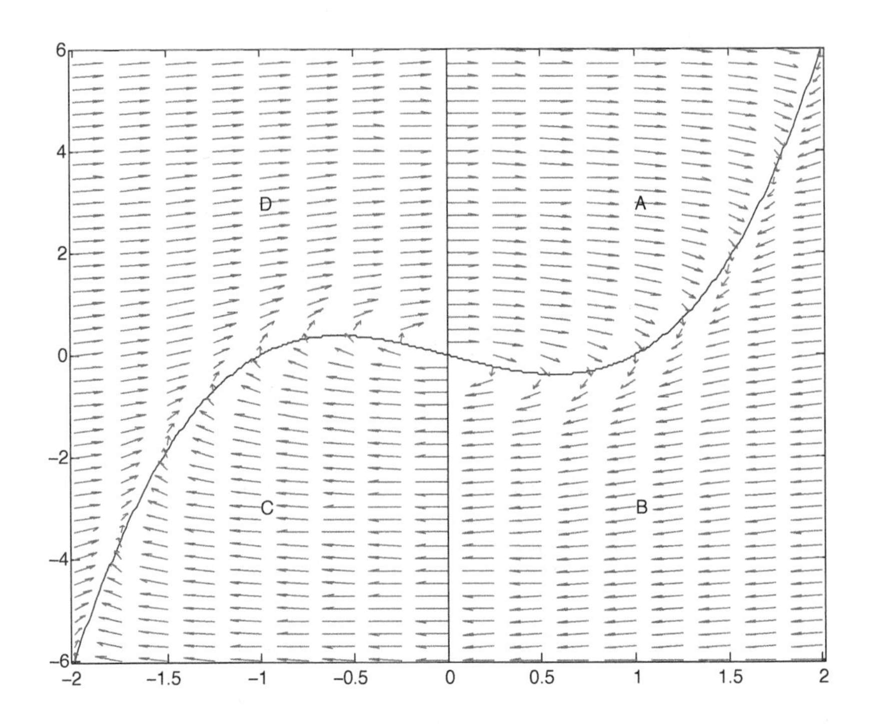

Figure 10.6: Regions A,B,C, and D.

Boundedness into the future.

We will prove:

Theorem 10.7.1. *Under the above hypotheses, there exists one periodic solution which is the ω limit set of all non-zero trajectories.*

The proof will occupy the rest of this section. vskip.1in Notice that the Lienard equations

$$\begin{aligned}\dot{x} &= y - F(x)\\ \dot{y} &= -x\end{aligned}$$

are unchanged under the map $(x, y) \mapsto (-x, -y)$. Let us divide the plane into four regions: Region A is between the positive y-axis and the curve $y = F(x)$, Region B is between the curve $y = F(x)$ and the negative y-axis, Region C is between the negatve y-axis and the curve $y = F(x)$ and region D is between the curve $y = F(x)$ and the positive y-axis, as in Figure 10.6. Starting on the positive y-axis, $x(t)$ is increasing, and after a small time, t_1,

$$\dot{y} \le -x(t_1) < 0$$

so y is steadily decreasing as long as the solution lies in region A, so the solution curve must eventually cross the curve $y = F(x)$ and enter region B. It can never go back across the curve $y = F(x)$ in the right hand plane to get back from region B to region A. In fact, once it leaves a small neighborhood of the curve, it can not return to that neighborhood.

Throughout regions D and A we have $\dot{x} > 0$, and $\dot{y} < 0$, so going back in time, any non-zero trajectory in A must have come from a point on the positive y-axis.

Since in region B we have $\dot{x} < 0$, the x-coordinate is decreasing. So by compactness, we will have $\dot{x} < -c < 0$ along the solution curve, so after a finite amount of time the trajectory will have to cross the negative y-axis at a point $(0, y_1)$, entering region C.

By symmetry of the equations, the trajectory then moves from C to D and then crosses the y-axis again at some point $(0, y_2)$. To prove the boundedness of all trajectories, it is enough to show that if y_0 is sufficiently large, then $y_2 < y_0$.

The map $\beta : y_0 \mapsto y_1$.

Let β be the map of $\mathbb{R}^+ \to \mathbb{R}^-$ obtained by following the trajectory from $(0, y_0)$ to $(0, y_1) := (0, \beta(y_0))$.

The trajectory though $(0, y)$ on the positive y-axis will be a periodic trajectory if and only if $\beta(y) = -y$. We shall prove that there is exactly one solution to this equation. What amounts to the same thing, we will show that there is exactly one y_0 for which

$$\frac{1}{2}y_0^2 = R(0, y_0) = R(0, y_1) = \frac{1}{2}y_1^2.$$

Let

$$\delta(y) := R((0, \beta(y)) - R((0, y)) = \int_0^{t_1(0,y)} \dot{R}(x(t), y(t))dt$$

where t_1 is the time it takes for the trajectory to go from $(0, y)$ to $(0, \beta(y))$ and the integral is along the trajectory. We want to show that $\delta(y)$ has a unique zero, and that $\delta(y)$ is negative for large y.

Let r be the unique value of y such that the trajectory through $(0, r)$ passes through the unique point $(0, z)$ where the curve $y = F(x)$ crosses the x-axis.

The following lemma will give us more than enough information to prove our theorem:

Lemma 10.7.1. *1. If $0 < y < r$ then $\delta(y) > 0$.*

2. For $y \geq r$ the function $\delta(y)$ is a monotone decreasing function of y.

3. $\delta(y) \to -\infty$ as $y \to \infty$.

Lemma 10.7.2. *1. If $0 < y < r$ then $\delta(y) > 0$.*

2. For $y \geq r$ the function $\delta(y)$ is a monotone decreasing function of y.

3. $\delta(y) \to -\infty$ as $y \to \infty$.

Along any trajectory we have

$$\dot{R} = x\dot{x} + y\dot{y} = xy - xF(x) - xy$$

so

$$\dot{R} = -xF(x).$$

Proof of 1. Let $0 < y \leq r$. The trajectory through $(0, y)$ can not cross the trajectory through $(0, r)$ and hence along this trajectory $x \leq z$ and $x = z$ at only one point and that only if if $y = r$. So $F \leq 0$ and is strictly negative except for that one possibility. So $\dot{R} > 0$ (except possibly at one point along one trajectory) proving 1.

Proof of 2. Let $r < y < y'$ and break up the trajectory going from $(0, y)$ to $(0, \beta(y))$ into three pieces. part a until it crosses the line $x = z$, part b from the first crossing until the second crossing and part c from the second crossing until it hits the negative y-axis. Do the same for the trajectory starting at $(0, y')$.

Along parts a and c (and similarly a' and c') we can use x as a parameter and

$$\frac{dR}{dx} = \frac{\dot{R}}{\dot{x}} = \frac{-xF(x)}{y - F(x)}.$$

We have

$$\int_a \dot{L}dt = \int_0^z \frac{-xF(x)}{y - F(x)}dx < \int_0^z \frac{-xF(x)}{y' - F(x)}dx = \int_{a'} \dot{R}dt$$

(we are now using y and y' as parameters along the trajectory) since $y' < y$ throughout this portion of the trajectories.

Similarly along c and c' we have $y < y'$ but x is decreasing so again

$$\int_c \dot{L}dt < \int_{c'} \dot{L}dt.$$

To compare the integrals along b and b' we break b up into three pieces: b_1 where b is above the horizontal line where b' first touches the line $x = z$, then the portion b_2 between these horizontal lines, and then b_3 the remaining portion of b.

Along all of b and b' we can use y as a coordinate and

$$\frac{dR}{dy} = \left(\frac{\partial R}{\partial x}\right)\left(\frac{\dot{x}}{\dot{y}}\right) + \frac{\partial R}{\partial y}$$

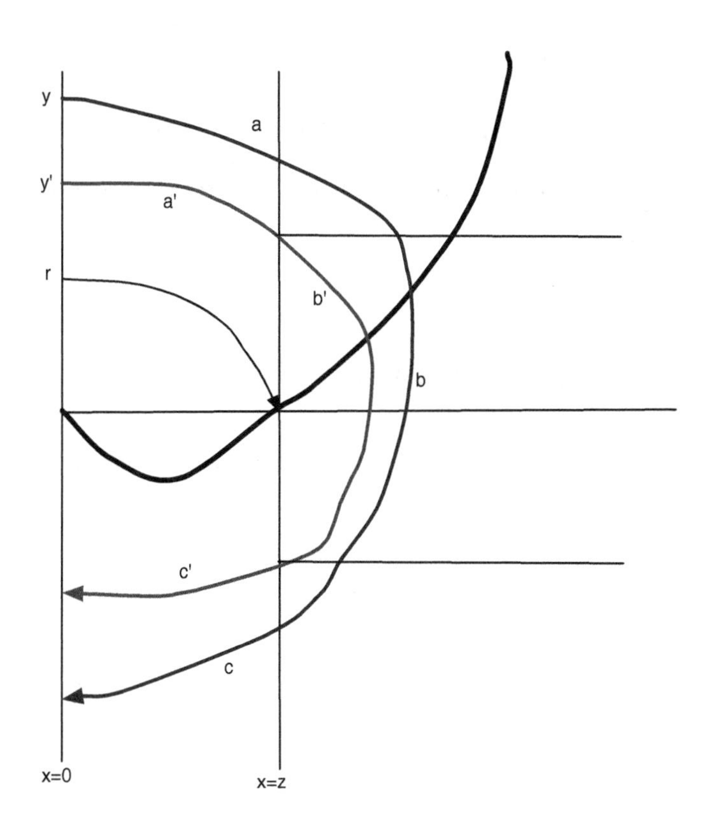

Figure 10.7: Diagram for the proof of part 2 of the lemma.

$$= x\left(\frac{y - F(x)}{-x}\right) + y = F(x)$$

Along all of b we know that $F \geq 0$ and y is decreasing so the contributions of b_1 and b_3 are negative. So we need only compare the integral over b_2 with the corresponding integral over b'.

But the x values along b_2 are greater than the corresponding x values along b' and F is monotone increasing for $x > z$ by hypothesis. So the integral over b_3 is $<$ the corresponding integral over b' proving part 2 of the lemma.

Proof of 3. As the initial point along the positive y-axis move up to infinity, all the x values on part b_2 of the curve b move to the right and $F \to \infty$. So the integral over this portion tends to $-\infty$. This completes the proof of the lemma and hence of the theorem. □

10.7.4 Relaxation oscillations.

We now want to consider the effect of the parameter μ on the equations

$$\begin{aligned} \dot{x} &= y - \mu F(x) \\ \dot{y} &= -x. \end{aligned}$$

Let us make the change of variable $w = \mu y$ so that the equations become

$$\begin{aligned} \dot{x} &= \mu(w - F(x)) \\ \dot{w} &= -\frac{1}{\mu}x. \end{aligned}$$

Suppose that μ is large. In the portion of the periodic motion where $w - F(x) > \epsilon > 0$ and $\mu\epsilon \gg 0$ the trajectory is practically horizontal by the second equation, and x is moving very fast to the right by the first equation. Once the trajectory gets within $O(\mu^{-2})$ of the curve $w = F(x)$ the right hand side of the first equation is small and there is a slow motion (initially vertically downward as we have seen) with x slowly decreasing. Eventually it will pull away from the curve $w = F(x)$ and x will rapidly decrease.

This is why this type of motion is called a "relaxation oscillation".

Balthasar van der Pol
Born: 27 Jan 1889 in Utrecht, The Netherlands
Died: 6 Oct 1959 in Wassenaar, The Netherlands.

Chapter 11

Lotka - Volterra.

The Lotka - Volterra predator prey equations were discovered independently by Alfred Lotka and by Vito Volterra in 1925-26. These equations have given rise to a vast literature, some of which we will sample in this chapter.

Here is how Volterra got to these equations: The number of predatory fishes immediately after WWI was much larger than before the war. The question as to why this was so was posed to the mathematician Volterra by his prospective son-in-law Ancona who was a marine biologist.

Much of the more recent results (in the second part of this chapter) are taken from the book [?] .

For a discussion of some of these issues at a level requiring less mathematics that we require in this chapter (and hence without some of the proofs) see the book *Evolutionary Dynamics* by Martin Nowak.

11.1 The original Lotka - Volterra equations.

Here is Volterra's solution to the problem posed to him by his prospective son-in-law:

Let x denote the density of prey fish and y denote the density of predator fish. Assume the equations

$$\begin{aligned} \dot{x} &= x(a-by) \\ \dot{y} &= y(-c+dx) \end{aligned} \tag{11.1}$$

where

$$a, b, c, d > 0.$$

The idea of the first equation is that in the absence of predators, the prey would grow at a constant rate a, but decreases linearly as a function of the density y of the predators. Similarly, in the absence of prey, the density of predators would decrease but the rate increases proportional to the density of the prey.

We are interested in solutions to these differential equations in the first quadrant

$$\mathbb{R}^2_+ = \{(x,y)|x \geq 0, y \geq 0\}.$$

11.1.1 The null-clines and the zeros.

The **null-clines** (where either $\dot{x} = 0$ or $\dot{y} = 0$ are zero) are the coordinate axes and the lines $y = a/b$ and $x = c/d$. The first quadrant is invariant . The origin is a saddle point.

The other point where the right hand side of (11.1) is zero is

$$\begin{pmatrix} \overline{x} \\ \overline{y} \end{pmatrix} := \begin{pmatrix} \frac{c}{d} \\ \frac{a}{b} \end{pmatrix}$$

where the linearized equation has matrix

$$\begin{pmatrix} 0 & -bc/d \\ da/b & 0 \end{pmatrix}$$

with purely imaginary eigenvalues $\pm i\sqrt{ac}$.

If we multiply the first equation of (11.1) by $(c - dx)/x$ and the second by $(a - by)/y$ and add we get

$$\left(\frac{c}{x} - d\right)\dot{x} + \left(\frac{a}{y} - b\right)\dot{y} = 0$$

or

$$\frac{d}{dt}[c \log x - dx + a \log y - by] = 0.$$

Let

$$H(x) := \overline{x}\log x - x, \quad G(y) := \overline{y}\log y - y,$$

and

$$V(x,y) := dH(x) + bG(y).$$

Then V is constant on flow lines.

Since

$$\frac{dH}{dx} = \frac{\overline{x}}{x} - 1, \quad \frac{d^2H}{dx^2} = -\frac{\overline{x}}{x^2} < 0$$

we see that H achieves a maximum at $\overline{x}$ and similarly G assumes a maximum at $\overline{y}$. Thus V has a unique maximum in the interior of the quadrant at the critical point. Thus the level curves of V, which are solution curves, are closed curves: all trajectories are periodic.

Here are some trajectories:

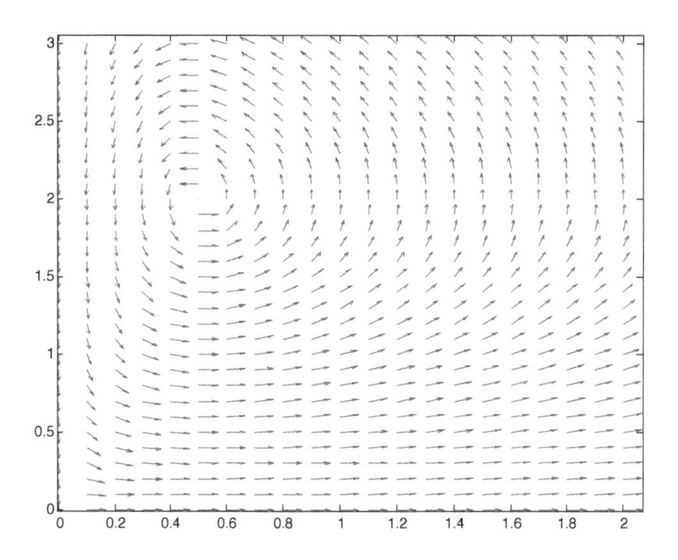

Figure 11.1: The Lotka -Volterra vector field with a=2, b=1, c=.25,d=1.

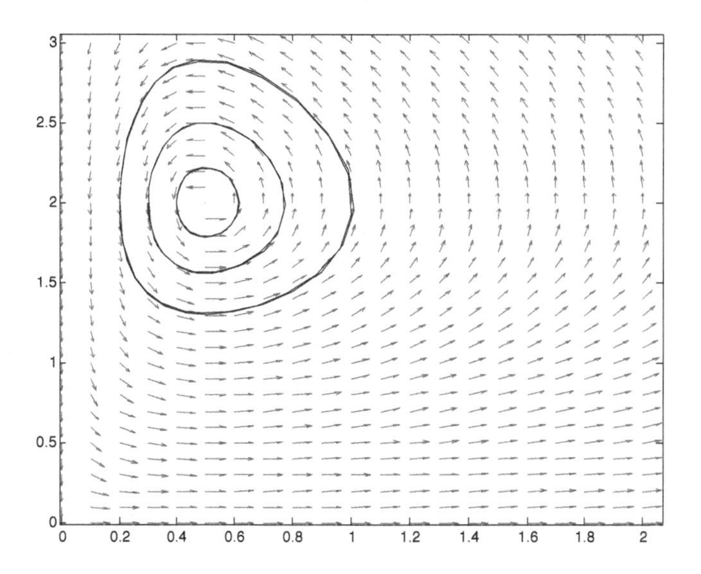

Figure 11.2: Some trajectories.

The fixed point as an average.

Suppose we are on a trajectory of period T, so $x(T) = x(0)$. From

$$\frac{d}{dt} \log x = \frac{\dot{x}}{x} = a - by$$

it follows by integration that

$$0 = \log x(T) - \log x(0) = aT - b \int_0^T y(t)dt$$

or

$$\frac{1}{T} \int_0^T y(t)dt = \overline{y}$$

and similarly

$$\frac{1}{T} \int_0^T x(t)dt = \overline{x}.$$

11.1.2 Volterra's explanation of why fishing decreases the number of predators.

Fishing reduces the rate of increase of the prey, so a is replaced by $a - k$ and increases the rate of decrease of the predator, so c is replaced by $c + m$, but does not change b or d - the interaction coefficients. So a/b is replaced by $(a - k)/b$ - the average number of predators is decreased by fishing and the average number of prey is increased. Stoppage of fishing increases the average number of predators and decreases average the number of prey.

A moral lesson.

If the prey are "pests" and the predators are their natural enemies, applying non-specific insecticides may actually increase the pest population.

11.2 A more realistic model.

Suppose we make the equations more realistic by adding self competition terms and so get the equations

$$\begin{aligned} \dot{x} &= x(a - ex - by) \\ \dot{y} &= y(-c + dx - fy) \end{aligned} \tag{11.2}$$

where

$$a, b, c, d, e, f > 0.$$

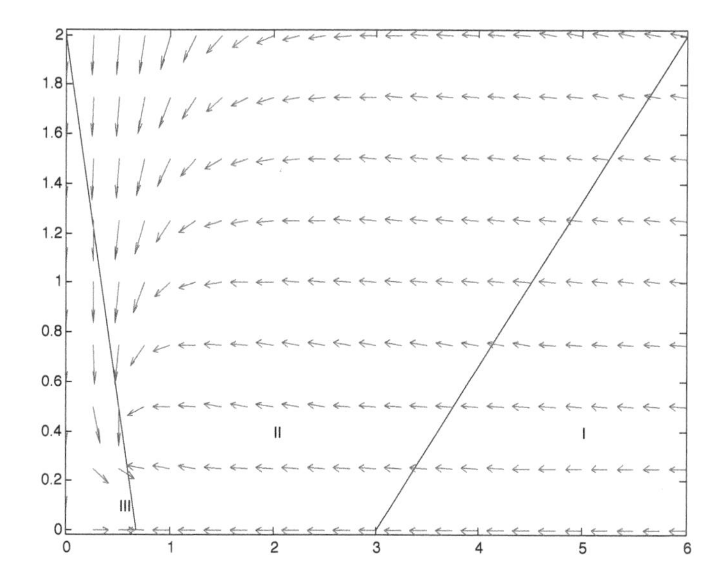

Figure 11.3: The null-clines do not intersect in the first quadrant, so the predators become extinct.

The first quadrant is still invariant, and there is an equilibrium point on the x-axis at $x = a/e$. There is no equilibrium point on the positive y-axis. The null-clines are now the axes and the two lines

$$ex + by = a, \quad \text{and} \quad dx - fy = c$$

the first with negative slope and the second with positive slope.

All hinges on whether or not they intersect in the first quadrant. If they don't, the quadrant is divided into three regions: in the region I to the right of the line $\dot{y} = 0$ of positive slope, we have $\dot{x} < 0$ so a trajectory starting in this region moves to the left, entering region II. It keeps moving to the left until it crosses the line $\dot{x} = 0$, entering region III, where it points down and to the right and heads toward the fixed point on the x-axis. The predators become extinct. The second alternative is that the null-clines intersect in the first quadrant, dividing it into four regions: In Figure 11.5 are some trajectories draw with Matlab's ode45. It looks as if trajectories (except those on the axes) are spiraling in to the zero of the vector field. Let's prove this: Label the fixed point as

$$\begin{pmatrix} \overline{x} \\ \overline{y} \end{pmatrix}.$$

With the same H, G and V as before, namely

$$H(x) := \overline{x} \log x - x, \quad G(y) := \overline{y} \log y - y,$$

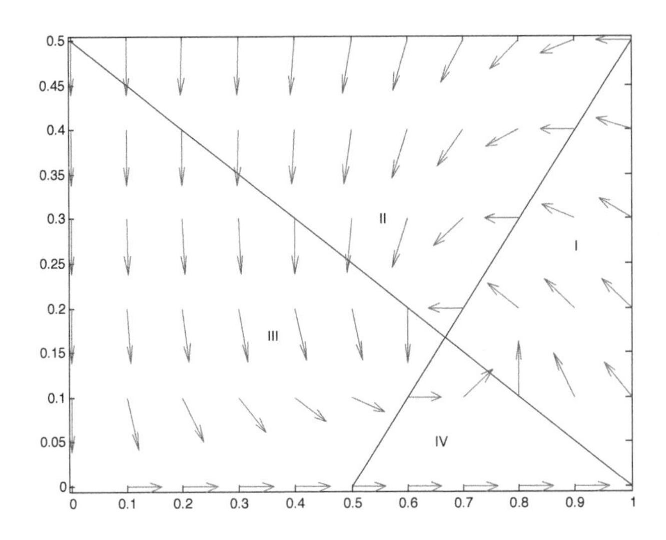

Figure 11.4: The null-clines intersect in the first quadrant, dividing it into four regions.

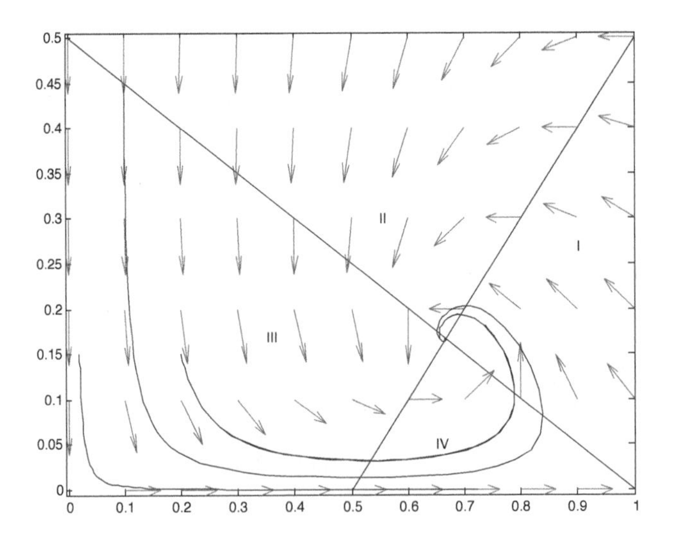

Figure 11.5: Some trajectories from ode45.

and

$$V(x,y) := dH(x) + bG(y)$$

we have

$$\dot{V} = \frac{\partial V}{\partial x}\dot{x} + \frac{\partial V}{\partial y}\dot{y}$$

$$= d(\frac{\overline{x}}{x} - 1)x(a - by - ex) + b(\frac{\overline{y}}{y} - 1)y(-c + dx - fy).$$

Write $a = e\overline{x} + b\overline{y}$ and $c = d\overline{x} - f\overline{y}$. We get

$$d(\overline{x} - x)(b\overline{y} + e\overline{x} - by - ex) + b(\overline{y} - y)(-d\overline{x} + f\overline{y} + dx - fy)$$

which simplifies giving

$$\dot{V}(x,y) = de(\overline{x} - x)^2 + bf(\overline{y} - y)^2.$$

This is non-negative, and strictly positive except at the equilibirum point. Hence V is steadily increasing along each orbit, which must then head to the equilibrium point. □

11.3 Competition between species.

If x and y denote the density of populations of two species competing for the same resources, then the rates of growth $\dot{x}/x$ and $\dot{y}/y$ will be decreasing functions of both x and y. The simplest assumption is that these decreases be linear which leads to the equations

$$\begin{aligned} \dot{x} &= x(a - bx - cy) \\ \dot{y} &= y(d - ex - fy) \end{aligned} \tag{11.3}$$

with

$$a, b, c, d, e, f > 0.$$

Again the first quadrant $\mathbb{R}^2_+$ is invariant. The x and y null-clines are given by the lines

$$\begin{aligned} a - bx - cy &= 0 \\ d - ex - fy &= 0 \end{aligned}$$

this time both of negative slope.

We will ignore the degenerate case where these lines are parallel. So we are left with two possibilities:

- The point of intersection does not lie in $\mathbb{R}^2_+$.
- The point of intersection does lie in $\mathbb{R}^2_+$.

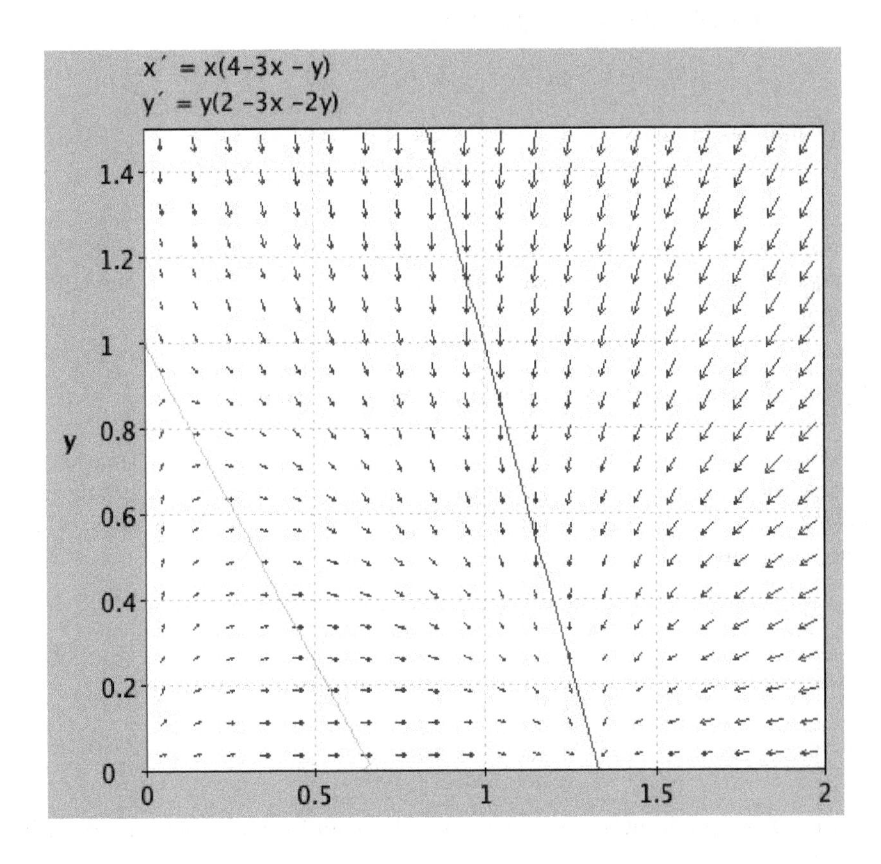

Figure 11.6: Case 1: y becomes extinct and x survives.

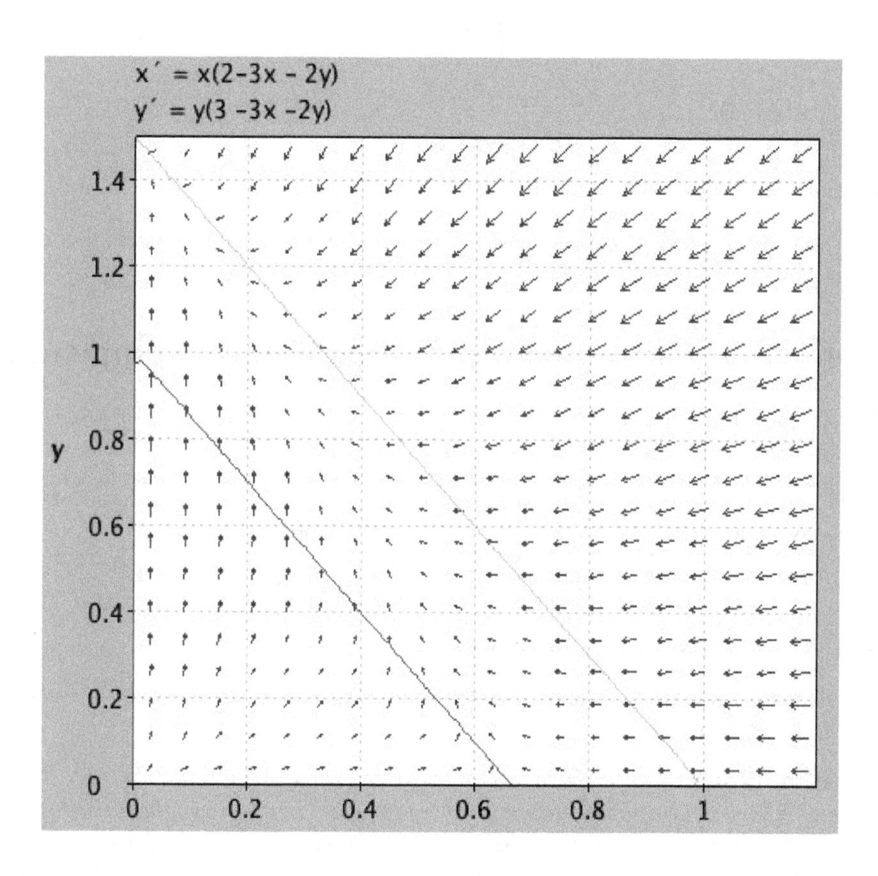

Figure 11.7: Case 2: x becomes extinct and y survives.

Each of these in turn leads to two possibilities:

If the null-clines do not intersect in the first quadrant then one or the other species will become extinct: If the two null-clines intersect in $\mathbb{R}^2$ the point of intersection is at

$$\overline{x} = \frac{af - cd}{bf - ce}, \quad \overline{y} = \frac{bd - ae}{bf - ce}$$

The Jacobian matrix at this point is

$$J = \begin{pmatrix} -b\overline{x} & -c\overline{x} \\ -e\overline{y} & -f\overline{y} \end{pmatrix}$$

with determinant

$$det(J) = \overline{xy}(bf - ce).$$

The eigenvalues are real. We can have (case 3)- a saddle, or (case 4) a sink.

In case 3 one or the other species dominates depending on the initial conditions:

11.4 The n-dimensional Lotka-Volterra equation.

In two dimensions, because of Poincaré-Bendixon, we can get more or less complete answers to the global behavior of flows.

We will now embark on the study of the higher dimensional version of the Lotka-Volterra equations where the answers are far less complete.

But we can say something:

11.4.1 A theorem of Liapounov.

We will use a theorem of Liapounov describing the ω-limit set in the presence of a "Liapounov function".

Theorem 11.4.1. *Let X be a vector field on some open set $O \subset \mathbb{R}^n$. Let $V : O \to \mathbb{R}$ be a continuously differentiable function. Let $t \mapsto x(t)$ be a trajectory of X. If the derivative $\dot{V}$ of the map $t \mapsto V(x(t))$ satisfies $\dot{V} \geq 0$ (for all t) then $\omega(x) \cap O$ is contained in the set where $XV = 0$.*

Remark. Along $x(t)$ we have $\dot{V}(t) = X(x(t))V$.

Proof. If $y \in \omega(x) \cap O$, there is a sequence $t_k \to \infty$ with $x(t_k) \to y$. Hence, by continuity, $(XV)(y) \geq 0$. If XV does not vanish at y, then $(XV)(y) > 0$ and we must show that this can not happen.

If $(XV)(y) > 0$, then for small positive values of t we would have $V(y(t)) > V(y)$. Now by hypothesis $V(x(s))$ is a monotone increasing function of s, and since $x(t_k) \to y$ and $t_k \to \infty$, for any s and sufficiently large k we have

$$V(x(s)) \leq V(x(t_k)) \to V(y).$$

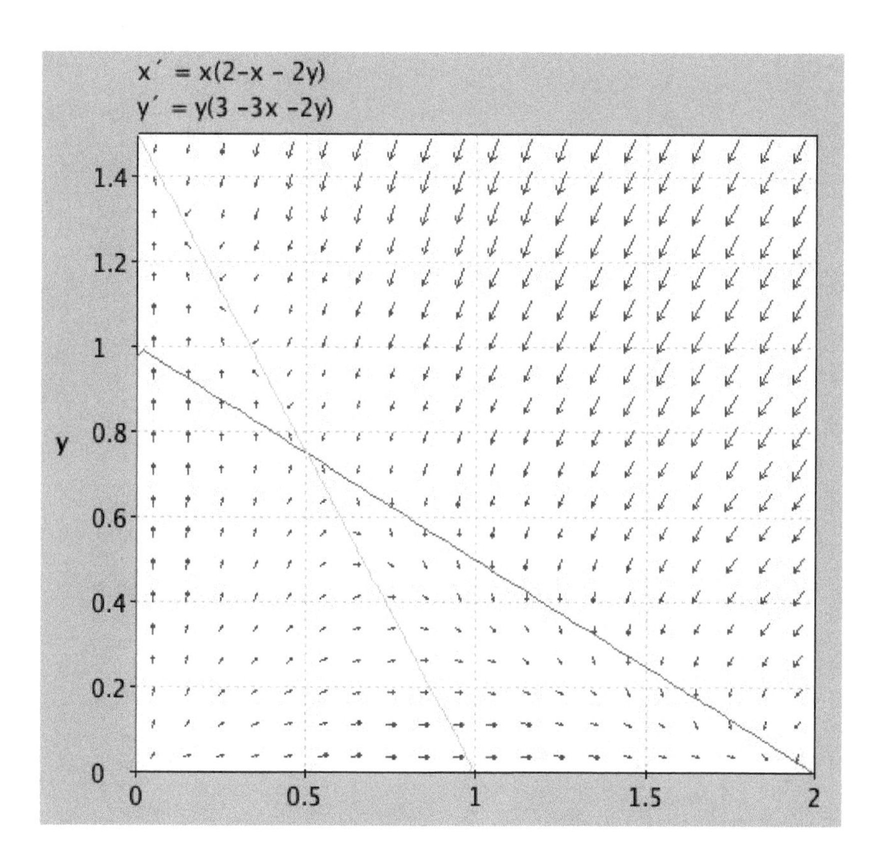

Figure 11.8: Case 3. One or the other species dominates depending on the initial conditions.

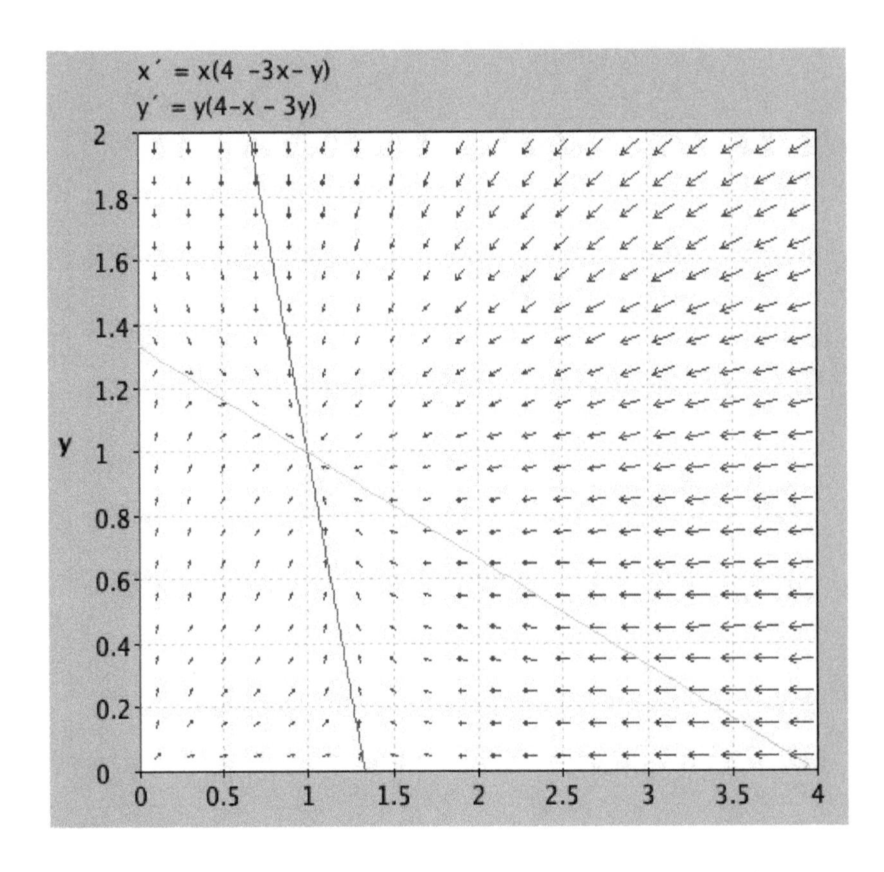

Figure 11.9: Case 4. Stable mututal coexistence.

So

$$V(x(s)) \leq V(y)$$

for all s.

Since $x(t_k) \to y$ we have (for small $t > 0$)

$$x(t_k + t) \to y(t)$$

and hence

$$V(x(t_k + t)) \to V(y(t)) > V(y),$$

a contradiction. □

With this in hand, we turn to the Lotka-Volterra equations for n species. These are:

$$\dot{x}_i = x_i \left(r_i + \sum_j a_{ij} x_j \right), \qquad i = 1, \ldots, n. \tag{11.4}$$

Here x_i denotes the density of the i-th species, r_i is its intrinsic growth (or decay) rate and the matrix $A = (a_{ij})$ is called the **interaction matrix**. .

Proposition 11.4.1. *The positive orthant and its faces are invariant under (11.4).*

Proof. If $x_i(0) = 0$, then $x_i(t) \equiv 0$ is a solution of the i-th equation, and hence by the uniqueness theorem of differential equations the only solution. So each of the faces $x_i = 0$ of the positive orthant $\mathbb{R}^n_+$ is invariant under the flow, and hence so is the (interior of) the positive orthant itself. □

The vector field given by the right hand side of the equations vanishes when

$$r_i + \sum_j a_{ij} x_j = 0, \tag{11.5}$$

and so zeros of the vector field in the positive orthant correspond to solutions of these equations with all $x_i > 0$.

Proposition 11.4.2. *An interior α or ω point implies a solution of (11.5) with all positive entries.*

Proof. To prove this, it is enough to show that if there is no solution to the above equation, then there is a "Liapounov function" V with $XV > 0$ everywhere, since Liapounov's theorem tells us that at any interior ω point we must have $XV = 0$ (and similarly for α points). To construct V, let L be the map $Lx = y$ where

$$y_i = r_i x_i + \sum_j a_{ij} x_j.$$

The image of the positive orthant is some convex cone C. The assumption is that C does not contain the origin. Since C is convex, there is a hyperplane separating it from the origin. Put another way, there is a vector c such that $c \cdot y > 0$ for all $y \in C$. Now define

$$V(x) := \sum_i c_i \log x_i$$

for all x in the positive orthant. Then

$$\dot{V} = \sum c_i \frac{\dot{x}_i}{x_i} = \sum c_i y_i > 0$$

at all points. □

In particular,

Corollary 11.4.1. *If there is a periodic solution in the positive orthant, there must also be a fixed point.*

11.4.2 Food chains.

A food chain is a system where the first species is prey for the second, the second is prey for the third , etc. up to the n-th which is at the top of the pyramid. Taking competition within each species into account, the differential equations are:

$$\begin{array}{rcl} \dot{x}_1 & = & x_1(r_1 - a_{11}x_1 - a_{12}x_2) \\ \dot{x}_2 & = & x_2(-r_2 + a_{21}x_1 - a_{22}x_2 - a_{23}x_3) \\ \vdots & \vdots & \vdots \\ \dot{x}_j & = & x_j(-r_j + a_{j,j-1}x_{j-1} - a_{jj}x_j - a_{j,j+1}x_{j+1}) \\ \vdots & \vdots & \vdots \\ \dot{x}_n & = & x_n(-r_n + a_{n,n-1}x_{n-1} - a_{nn}x_n) \end{array} \tag{11.6}$$

with all the r_i and a_{ij} positive.

Theorem 11.4.2. *If the food chain equations have an interior rest point*

$$p = \begin{pmatrix} p_1 \\ \vdots \\ p_n \end{pmatrix},$$

i.e. a point p where the right hand side of the food chain equations vanish, then p is globally stable in the sense that all orbits in the interior of the positive orthant converge to p.

The proof will consist of constructing a Liapounov function of the form

$$V(x) = \sum c_i(x_i - p_i \log x_i)$$

for suitably chosen constants c_i.

Proof. Let $V(x) = \sum c_i(x_i - p_i \log x_i)$. Then

$$\dot{V}(x) = \sum c_i \left(\dot{x}_i - p_i \frac{\dot{x}_i}{x_i} \right).$$

If we write the food chain equations as $\dot{x}_i = x_i w_i$ this becomes

$$\dot{V}(x) = \sum c_i (x_i - p_i) w_i.$$

By assumption, the w_i vanish at p. So, for example, $r_1 = a_{11}p_1 + a_{22}p_2$ hence

$$w_1 = r_1 - a_{11}x_1 - a_{12}x_2 = a_{11}(p_1 - x_1) + a_{12}(p_2 - x_2).$$

More generally we have

$$\begin{aligned}
w_1 &= a_{11}(p_1 - x_1) + a_{12}(p_2 - x_2) \\
w_2 &= a_{21}(x_1 - p_1) - a_{22}(x_2 - p_2) - a_{23}(x_3 - p_3) \\
\vdots &\ \vdots\ \vdots \\
w_n &= a_{n,n-1}(x_{n-1} - p_{n-1}) - a_{nn}(x_n - p_n).
\end{aligned}$$

So if we set $y_i := x_i - p_i$ we get

$$\dot{V} = -c_1 a_{11} y_1^2 - y_1 y_2 c_1 a_{12} - c_2 a_{22} y_2^2 + c_2 a_{21} y_1 y_2 c_2 - a_{23} y_2 y_3$$

$$-c_3 a_{33} y_3^2 + y_2 y_3 c_3 a_{32} - c_3 a_{34} y_3 y_4 + \cdots$$

Rearranging terms on the right hand side this becomes

$$\dot{V} = -\sum_{j=1}^{n} c_j a_{jj} y_j^2 + \sum_{j=1}^{n-1} y_j y_{j+1}(-c_j a_{j,j+1} + c_{j+1} a_{j+1,j}).$$

Since all the $a_{j,j+1}$ and $a_{j+1,j}$ are positive, we can choose $c_j > 0$ recursively such that $-c_j a_{j,j+1} + c_{j+1} a_{j+1,j} = 0$, i.e. $c_{j+1}/c_j = a_{j,j+1}/a_{j+1,j}$. Then the second summand above vanishes, and we have

$$\dot{V} = -\sum_{j=1}^{n} c_j a_{jj} y_j^2 \leq 0$$

with strict inequality unless all the $y_i = 0$.

By Liapounov's theorem, the ω limit of every orbit in the interior of the postive orthant is p. □

11.5 Replicator dynamics and evolutionary stable strategies.

11.5.1 The replicator equation.

We will let $\overline{S_n}$ denote the simplex consisting of all

$$x = \begin{pmatrix} x_1 \\ \vdots \\ x_n \end{pmatrix}$$

with $x_i \geq 0$ and $\sum_i x_i = 1$. We want to think of a population as being divided into n types $E_1, \ldots, E_n$ and of x_i as the frequency of the i-th type E_i. The "fitness" f_i of E_i will be a function of these frequencies, i.e. of x. If the population is very large and the generations blend continuously into each other, we may assume that $x(t)$ is differentiable function of t. The rate of increase of $\dot{x}_i/x_i$ is a measure of the evolutionary success of type E_i. The basic tenet of Darwinism says that we may express this success as the difference between $f_i(x)$ and the average fitness $\overline{f}(x) := \sum x_i f_i(x)$. We obtain the **replicator equation**

$$\dot{x}_i = x_i(f_i(x) - \overline{f}(x)). \tag{11.7}$$

If we set $S(x) := x_1 + \cdots + x_n$, then summing the above equations gives the equation

$$\dot{S} = (1 - S)\overline{f}.$$

The (unique) solution of this equation with $S(0) = 1$ is $S(t) \equiv 1$. So the set $\overline{S_n}$ is preserved by the flow. Also, if $x_i(0) = 0$ for some i, then $x_i(t) \equiv 0$. Thus the faces of $\overline{S_n}$ are preserved, and hence so is the open simplex

$$S_n := \{x \in \overline{S_n} | x_i > 0 \ \forall i\}.$$

11.5.2 Linear fitness.

For most of the rest of this chapter, I will assume that the fitnesses are linear functions of x, i.e. there is a matrix A such that $f_i(x) = (Ax)_i$ - the i-th component of Ax. The replicator equations are then cubic equations in x:

$$\dot{x}_i = x_i \left((Ax)_i - x \cdot Ax \right).$$

We shall see that a change of variables will carry the orbits of the replicator equation with linear fitness functions to the orbits of the Lotka-Volterra equations (which are quadratic) in one fewer variables, a result due to Hofbauer.

These equations are equivalent in the above sense, but some notions are easier to formulate and understand in one setting, and some in the other.

11.5.3 Hofbauer's equivalence theorem.

Some preliminaries to the equivalence theorem.

Let us go back temporarily to the general replicator equation (11.7): If we add a function $h = h(x)$ to all the f_i, this has the effect of replacing $\overline{f}$ by $\overline{f} + h$ since $\sum x_i h = h$ as $\sum x_i = 1$. Thus the right hand side of the above equation is unchanged. So adding a common function to all the f_i does not change the replicator equation.

In case $f_i = (Ax)_i$, if we add a constant c to all the entries in the j-th column of A, this has the effect of adding the function cx_i to all the f_i, so does not change the replicator equation. In particular, this means that (by subtracting offthe entry in the last row from each column) we can assume that our matrix A has all entries in the bottom row zero, without changing the replicator equation.

Here is another useful fact about the general replicator equation (11.7): We have

$$\begin{aligned}\left(\frac{x_i}{x_j}\right)^{\cdot} &= \frac{\dot{x}_i x_j - x_i \dot{x}_j}{x_j^2} \\ &= \frac{(f_i - \overline{f})x_i x_j - (f_j - \overline{f})x_i x_j}{x_j^2} \\ &= \frac{(f_i - f_j)x_i x_j}{x_j^2}\end{aligned}$$

so

$$\left(\frac{x_i}{x_j}\right)^{\cdot} = \left(\frac{x_i}{x_j}\right)(f_i(x) - f_j(x)). \tag{11.8}$$

Consider the map of the set $\{y \in \mathbb{R}^n_+ | y_n = 1\}$ onto the set

$$\hat{S}_n := \{x \in \overline{S_n} | x_n > 0\}$$

given by

$$x_i = \frac{y_i}{\sum_j y_j}, \quad i = 1, \ldots, n.$$

The inverse map $x \mapsto y$ is given by

$$y_i = \frac{y_i}{y_n} = \frac{x_i}{x_n}.$$

If x satisfies the general replicator equation then

$$\dot{y}_i = y_i(f_i(x) - f_n(x))$$

by (11.8).

Hofbauer's theorem.

Now suppose that $f_i(x) = (Ax)_i$ and we have chosen the matrix A to have its bottom row all zero (which we can do without changing the equations). Then $f_n(x) \equiv 0$ and the preceding equations become

$$\dot{y}_i = y_i \sum_{j=1}^{n} a_{ij}x_j = y_i \left(\sum_{j=1}^{n-1} a_{ij}y_j \right) x_n.$$

The positive factor x_n affects the speed with which the trajectories are traveled, but not the shape of the trajectories themselves.

In other words, the trajectories in y space are given by

$$\dot{y}_i = y_i \left(a_{in} + \sum_{j=1}^{n-1} a_{ij}y_j \right) \qquad i = 1, \ldots n-1.$$

If we consider a general matrix (and then modify it to get the bottom row zero) we have proved

Theorem 11.5.1. **[Hofbauer.]** *The differentiable invertible map $x \mapsto y$ given above maps the orbits of the replicator equation with linear fitness $f_i = (Ax)_i$ onto the orbits of the Lotka-Volterra equation*

$$\dot{y}_i = y_i \left(r_i + \sum_{j=1}^{n-1} b_{ij}y_j \right) \qquad i = 1, \ldots n-1$$

where

$$r_i = a_{in} - a_{nn} \quad \text{and} \quad b_{ij} = a_{ij} - a_{nj}.$$

The steps in this passage from replicator equations to LV are reversible.

11.5.4 Nash equilibria.

Let us go back he replicator equations with linear fitnesses. The right hand side of the equation $\dot{x}_i = x_i((Ax)_i - x \cdot Ax)$ vanishes if and only if all the $(Ax)_i$ are equal (in which case they all equal $x \cdot Ax$). So the conditions for such a rest point are the equations

$$(Ax)_1 = \cdots = (Ax)_n, \quad x_1 + \cdots x_n = 1, \quad x_i > 0 \; \forall \; i,$$

n equations in n unknowns, which, therefore, will generically have one or no solutions. These equations are related to certain concepts and equations in game theory: For a given matrix A a point $p \in \overline{S_n}$ is called a **Nash equilibrium** if

$$x \cdot Ap \leq p \cdot Ap \quad \forall x \in \overline{S_n}.$$

If p is a Nash equilibrium, then taking $x = e_i$ (the i-th unit vector) in the above inequality gives

$$(Ap)_i \leq p \cdot Ap.$$

Multiplying by p_i and summing i gives us back $p \cdot Ap$. So we can not have strict inequality for any i for which $p_i > 0$. We must have

$$(Ap)_i = p \cdot Ap \qquad \forall\ i \text{ for which } p_i > 0.$$

Interior Nash equilibria are rest states of the replicator equation.

In particular, if $p \in S_n$ (the interior of the simplex) - so that all the $p_i > 0$) then we have

$$(Ap)_i = p \cdot Ap \qquad \forall\ i,$$

and so the right hand side of the replicator equation vanishes at p.

More generally, if we consider the face of $\overline{S_n}$ spanned by those e_i for which $p_i > 0$, we see that p is a rest point of the replicator equation (restricted ot the interior of that face).

We know that for the Lotka-Volterra equations the existence of an interior ω limit point of any orbit implies the existence of an interior rest point, and we know that the replicator orbits have the same structure as the LV orbits. So we have proved:

Theorem 11.5.2. *If $p \in \overline{S_n}$ is a Nash equilibrium of A then it is a rest point for the associated replicator equation.*

We also have:

Theorem 11.5.3. *If $x(t) \to p \in S_n$ as $t \to \infty$ for some orbit then it is a Nash equilibrium.*

Proof. Suppose that p is *not* a Nash equilibrium. Then for some i we have $(Ap)_i - p \cdot Sp > 2\epsilon > 0$. Since $x(t) \to p$, this means that for sufficiently large t we have $\dot{x}_i/x_i > \epsilon$ which is clearly impossible. □

11.6 Evolutionary stable states.

A point $p \in S_n$ is called an **evolutionary stable state** if

$$p \cdot Ax > x \cdot Ax \quad \forall\ x \neq p, \quad x \in S_n.$$

Theorem 11.6.1. **[Zeeman.]** *If p is an evolutionary stable state then every orbit of the associated replicator equation in the open simplex S_n converges to p.*

For the proof of the theorem we will use the inequality

$$\log x \leq x - 1$$

for $x > 0$ with strict inequality when $x \neq 1$.

To prove this inequality observe that both sides are equal when $x = 1$. For $x > 1$ the derivative of the right hand side is 1 while the derivative of the left hand side is $1/x < 1$, so the right hand side is increasing faster. For $x < 1$ we have $1/x > 1$ so the left hand side is increasing faster, and so is strictly below $x - 1$. □

A very important consequence of this simple inequality is the inequality

$$\sum p_i \log x_i \leq \sum p_i \log p_i \quad \text{with strict inequality unless } x_i = p_i \quad \text{for all } \ i. \tag{11.9}$$

. To prove this:

$$\sum p_i \log x_i - \sum p_i \log p_i = \sum p_i \log \frac{x_i}{p_i} \leq \sum p_i \left(\frac{x_i}{p_i} - 1\right)$$

$$= \sum x_i - \sum p_i = 1 - 1 = 0.$$

The inequality becomes strict if any $x_i \neq p_i$. □

So the function $V(x) = \sum p_i \log x_i$ achieves its maximum at p. We shall show that if p is an evolutionary stable state then V is a Liapounov function for the associated replicator equation.

Proof of Zeeman's theorem. Indeed,

$$\dot{V} = \sum p_i \frac{\dot{x}_i}{x_i} = \sum p_i \left((Ax)_i - x \cdot Ax\right) = p \cdot Ax - x \cdot Ax > 0$$

if $x \neq p$. □

Relation to information theory.

The function $\text{Ent}(x) := -\sum x_i \log x_i$ (known as the **entropy**) plays a key role in thermodynamics, statistical mechanics, and information theory. As a diversion, I will spend the rest of this chapter trying to explain why and how this enters into communication theory.

11.7 Entropy and communication.

11.7.1 Codes.

We use the following notations. W will be a set of "words", $W = \{a, b, c, d, \dots\}$. The number of elements in W will be denoted by N. A **message** is just a concatenation of words, i.e. a string of elements of W.

Σ will denote an alphabet of D symbols (usually $D = 2$ and the symbols are 0 and 1). A **code** or an **encoding** is a map from W to strings onΣ . It then extends by concatenation to messages.

Example.

$$W = \{a, b, c, d\} \quad \phi: \; a \mapsto 0, \; b \mapsto 111, \; c \mapsto 110, \; d \mapsto 101.$$

Then

$$\phi(aba) = 01110.$$

11.7.2 Uniquely decipherable codes and instantaneous codes.

A code, ϕ, is called **uniquely decipherable** (UD) if any string S on Σ has at most one preimage under ϕ. A code is called **instantaneous** (INS) if no $\phi(w)$ occurs as a prefix of the code for some other word. Clearly every instantaneous code can be uniquely deciphered, each word as it arrives. Hence

$$\text{UD} \supset \text{INS}.$$

Example. $W = \{x, y\}, \Sigma = \{0, 1\}, \phi: \; x \mapsto 0, \; y \mapsto 01$. Then

$$00010101$$

deciphers from the end as $xxyyy$. But we needed to wait until the end of the message to decode. If the last digit had been a 0 instead of a 1, it would have decoded as $xxyyxx$. So the inclusion is strict.

11.7.3 The expected length of a code.

We let $|S|$ denote the length (number of elements in) a string S. Suppose that the messages sent are all such that each word w occurs with a relative frequency $f(w)$, so we think of f as a probability measure on W. So now $f(w)$ denotes frequency rather than fitness.

Then the expectation

$$E(|\phi(w)|) = E_f(|\phi(w)|) = \sum f(w)|\phi(w)|$$

is the "average length of the encoding ϕ". We would like to make this as small as possible.

We define

$$\text{Ent}(f) = E(-\log f) = -\sum_w f(w) \log f(w)$$

as before.

11.7.4 Shannon's "first theorem".

Theorem 11.7.1. *For any UD code, ϕ we have*

$$E(|\phi(w)|) \geq \frac{\mathrm{Ent}(f)}{\log D}. \tag{11.10}$$

There exists an INS code ϕ such that

$$E(|\phi(w)|) \leq \frac{\mathrm{Ent}(f)}{\log D} + 1. \tag{11.11}$$

McMillan's inequality.

I will prove (11.10) by first proving *McMillan's inequality*: Let $\ell_1, \ldots, \ell_N$ be the code word lengths of a UD code. Then McMillan's inequality says that

$$\sum_1^N D^{-\ell_i} \leq 1. \tag{11.12}$$

The proof of McMillan's inequality will be by the method of generating functions:

For any integer r we have

$$\left(D^{-\ell_1} + D^{-\ell_2} + \cdots + D^{-\ell_N}\right)^r = \sum_1^{r\ell} b_i D^{-i}$$

where $\ell = \max \ell_j$ and where b_i denotes the number of ways that a string of length i can be constructed by concatenating r code words. Now there are D^i strings of length i in all. If the code is uniquely decipherable then there can't be more than D^i messages whose code is a string of length i. Hence

$$b_i \leq D^i$$

and plugging into the preceding equality gives

$$\left(D^{-\ell_1} + D^{-\ell_2} + \cdots + D^{-\ell_N}\right)^r \leq r\ell.$$

Hence

$$D^{-\ell_1} + D^{-\ell_2} + \cdots + D^{-\ell_N} \leq \ell^{1/r} r^{1/r} \to 1$$

as $r \to \infty$. This proves McMillan's inequality (11.12).

Now to the proof of Shannon's inequality (11.10): Set

$$q_i = \frac{D^{-\ell_i}}{D^{-\ell_1} + D^{-\ell_2} + \cdots + D^{-\ell_N}}$$

so that

$$\sum q_i = \sum f_i = 1$$

where $f_i = f(w_i)$ is the frequency of the i–th word. We have proved that

$$\mathrm{Ent}(f) = -\sum f_i \log f_i \leq -\sum f_i \log q_i = \log D \sum f_i \ell_i + \log(\sum_1^N D^{-\ell_i}).$$

But $(\sum_1^N D^{-\ell_i}) \leq 1$ and hence its logarithm is negative. We conclude that

$$\mathrm{Ent}(f) \leq \log D \times E(|\phi(w)|)$$

which is just (11.10). □

Krafts lemma.

We now show that there exists an instantaneous code satisfying (11.11). For this we need Kraft's lemma

Lemma 11.7.1. *For any ℓ_i satisfying (11.12) there exists an instantaneous code whose word lengths are ℓ_i.*

Proof. Write (11.12) as

$$\sum_1^\ell n_j D^{-j} \leq 1$$

where n_i is the number of ℓ_i which are equal to j. Multiply through by D^ℓ and move terms to the other side so the inequality becomes

$$n_\ell \leq D^\ell - n_1 D^{\ell-1} - \cdots - n_{\ell-1} D.$$

Now the n_ℓ which occurs on the left of the inequality is a non-negative (actually positive) integer. So we certainly have the inequality

$$0 \leq D^\ell - n_1 D^{\ell-1} - \cdots - n_{\ell-1} D.$$

Dividing by D and bringing $n_{\ell-1}$ over to the other side gives

$$n_{\ell-1} \leq D^{\ell-1} - n_1 D^{\ell-2} - \cdots - n_{\ell-2} D.$$

So proceding in this way we get the string of inequalities

$$\begin{array}{rcl} n_\ell & \leq & D^\ell - n_1 D^{\ell-1} - \cdots - n_{\ell-1} D \\ n_{\ell-1} & \leq & D^{\ell-1} - n_1 D^{\ell-2} - \cdots - n_{\ell-2} D \\ n_{\ell-2} & \leq & D^{\ell-2} - n_1 D^{\ell-3} - \cdots - n_{\ell-3} D \\ & \vdots & \\ n_3 & \leq & D^3 - n_1 D^2 - n_2 D \\ n_2 & \leq & D^2 - n_1 D \\ n_1 & \leq & D. \end{array}$$

Let us read these inequalities in reverse order. The last inequality says that we can encode n_1 words each by a single letter from the alphabetΣ , with $D - n_1$ letters left over to serve as prefixes of code words. The next to last inequality says that we can encode n_2 words as two letter code words using the $D - n_1$ letters as first letters and choosing from the D letters as second letters, $D^2 - n_1 D = (D - n_1)D$ possibilities in all. This leave $D^2 - n_1 D - n_2$ possible prefixes for three letter words, and the third from last inequality says that we have enough room to encode n_3 words as three letter code words. Proceeding in this way back up to the top proves Kraft's lemma. □

Now to the proof of the second assertion in Shannon's theorem. Choose word lengths ℓ_i to be the smallest integers satisfying

$$f_i^{-1} \le D^{\ell_i}.$$

This is equivalent to

$$\ell_i \log D \ge -\log f_i$$

and

$$\ell_i \le 1 - \frac{\log f_i}{\log D}$$

since we have chosen ℓ_i as small as possible. But

$$\sum D^{-\ell_i} \le \sum f_i = 1$$

so (11.12) is satisfied, and we can find an instantaneous code with the word lengths ℓ_i. For this code we have

$$\sum f_i \ell_i \le \sum f_i \left(1 - \frac{\log f_i}{\log D}\right) = 1 + \frac{\text{Ent}(f)}{\log D}. \qquad \square$$

A good book on the subject of this section is [Welsh].

Here are photographs of Lotka and Volterra

Alfred James **Lotka** (1880 – 1949)

Vito Volterra

1860 - 1940

Chapter 12

Symbolic dynamics.

We have already seen several examples where a dynamical system is conjugate to the dynamical system consisting of a "shift" on sequences of symbols. It is time to pin this idea down with some formal definitions.

Definition. A **discrete compact dynamical system** (M, F) consists of a compact metric space M together with a continuous map $F : M \to M$. If F is a homeomorphism then (M, F) is said to be an **invertible** dynamical system.

If (M, F) and (N, G) are compact discrete dynamical systems then a map $\phi : M \to N$ is called a **homomorphism** if

- ϕ is continuous, and
- $G \circ \phi = \phi \circ F$.

If the homomorphism ϕ is surjective it is called a **factor**. If ϕ a homeomorphism then it is called a **conjugacy**.

For the purposes of this chapter we will only be considering compact discrete situations, so shall drop these two words.

12.1 Sequence spaces.

Let $\mathcal{A}$ be a finite set called an "alphabet". The set $\mathcal{A}^{\mathbb{Z}}$ consists of all bi-infinite sequences $x = \cdots x_{-2}, x_{-1}, x_0, x_1, x_2, x_3, \cdots$. On this space let us put the metric $d(x, x) = 0$ and, if $x \neq y$ then

$$d(x, y) = 2^{-k} \text{ where } k = \max_i \, [x_{-i}, x_i] = [y_{-i}, y_i].$$

Here we use the notation $[x_k, x_\ell]$ to denote the "block"

$$[x_k, x_\ell] = x_k x_{k+1} \cdots x_\ell$$

from k to ℓ occurring in x. (This makes sense if $k \leq \ell$. If $\ell < k$ we adopt the convention that $[x_k, x_\ell]$ is the empty word.) Thus the elements x and y are close in this metric if they agree on a large central block. So a sequence of points $\{x^n\}$ converges if and only if, given any fixed k and ℓ, the $[x_k^n, x_\ell^n]$ eventually agree for large n.

From this characterization of convergence, it is easy to see that the space $\mathcal{A}^{\mathbb{Z}}$ is sequentially compact: Let x^n be a sequence of points of $\mathcal{A}^{\mathbb{Z}}$, We must find a convergent subsequence. The method is Cantor diagonalization: Since $\mathcal{A}$ is finite we may find an infinite subsequence n_i of the n such that all the $x_0^{n_i}$ are equal. Infinitely many elements from this subsequence must also agree at the positions -1 and 1 since there are only finitely many possible choices of entries. In other words, we may choose a subsequence n_{i_j} of our subsequence such that all the $[x_{-1}^{n_{i_j}}, x_1^{n_{i_j}}]$ are equal. We then choose an infinite subsequence of this subsubsequence such that all the $[x_{-3}, x_3]$ are equal. And so on. We then pick an element N_1 from our first subsequence, an element $N_2 > N_1$ from our subsubsequence, an element $N_3 > N_2$ from our subsubsubsequence etc. By construction we have produced an infinite subsequence which converges.

12.1.1 Exclusions.

In the examples we studied, we did not allow all sequences, but rather excluded certain types. Let us formalize this. By a **word** from the alphabet $\mathcal{A}$ we simply mean a finite string of letters of $\mathcal{A}$. Let $\mathcal{F}$ be a set of words. Let

$$X_{\mathcal{F}} = \{x \in \mathcal{A}^{\mathbb{Z}} | [x_k, x_\ell] \notin \mathcal{F}\}$$

for any k and ℓ. In other words, $X_{\mathcal{F}}$ consists of those sequences x for which no word of $\mathcal{F}$ ever occurs as a block in x.

12.1.2 Shifts.

From our characterization of convergence (as eventual agreement on any block) it is clear that $X_{\mathcal{F}}$ is a closed subset of $\mathcal{A}^{\mathbb{Z}}$ and hence compact. It is also clear that $X_{\mathcal{F}}$ is mapped into itself by the **shift map**

$$\sigma : \mathcal{A}^{\mathbb{Z}} \to \mathcal{A}^{\mathbb{Z}}, \quad (\sigma x)_k := x_{k+1}.$$

It is also clear that σ is continuous. By abuse of language we may continue to denote the restriction of σ to $X_{\mathcal{F}}$ by σ although we may also use the notation σ_X for this restriction. A dynamical system of the form (X, σ_X) where $X = X_{\mathcal{F}}$ is called a **shift** dynamical system.

Example: The full shift on two letters and the Baker's transformation.

Consider the alphabet $\mathcal{A} = \{0, 1\}$ and the full sequence space $\mathcal{A}^{\mathbb{Z}}$. A point of $\mathcal{A}^{\mathbb{Z}}$ looks like

$$\cdots x_{-4}x_{-3}x_{-2}x_{-1}x_0x_1x_2x_3x_4x_5 \cdots .$$

I am going to rewrite this point by putting a "." before the x_0 and setting $y_i = x_{-i}$ for $i > 0$. I will then think of $.x_0x_1x_2x_3\ldots$ as the binary expansion of a point x in the interval $[0,1]$ and of $.y_1y_2y_3\ldots$ as the binary expansion of a point y in the interval $[0,1]$. In this way we have a map

$$h : \mathcal{A}^{\mathbb{Z}} \to [0,1] \times [0,1].$$

So h maps our sequence space onto the unit square and is easily seen to be continuous .

Let us examine the effect of the shift **Sh** on the unit square. In other words, we are looking for a transformation **b** on the unit square such that $\mathbf{b} \circ h = h \circ \mathbf{Sh}$. If $x_0 = 0$, the effect of **Sh** on x is to replace it by $2x$, while at the same time to replace y by $\frac{1}{2}y$. If $x_0 = 1$, the effect on x is to replace it by $2x - 1$ while the effect on y is to replace it by $\frac{1}{2}y + \frac{1}{2}$. In other words, the transformation **b**, known as the "bakers transformation" is the composition $\mathbf{b} = \mathbf{c} \circ \mathbf{sq}$ where **sq** squashes the square by multiplying x by 2 and y by $\frac{1}{2}$, then cutting the right hand rectangle $[1,2] \times [0, \frac{1}{2}]$ and placing it on top of the rectangle $[0,1] \times [0, \frac{1}{2}]$. The idea is that in kneading dough, the first step is to squash down and flatten out the dough and then cut it and reassemble it. Then continue the process.

It is easy to check that the shift map is chaotic, and it follows from Prop. 4.3.3 that the Baker's transformation is chaotic.

12.1.3 Homomorphisms between shifts are sliding block codes.

Suppose that (X,σ_X) with $X = X_{\mathcal{F}} \subset \mathcal{A}^{\mathbb{Z}}$ and (Y,σ_Y) with $Y = Y_{\mathcal{G}} \subset \mathcal{B}^{\mathbb{Z}}$ are shift dynamical systems. What does a homomorphism $\phi : X \to Y$ look like? For each $b \in \mathcal{B}$, let

$$C_0(b) = \{y \in Y | y_0 = b\}.$$

(The letter C is used to denote the word "cylinder" and the subscript 0 denotes that we are constructing the so called cylinder set obtained by specifying that the value of y at the "base" 0.) The sets $C_0(b)$ are closed, hence compact, and distinct. The finitely many sets $\phi^{-1}(C_0(b))$ are therefore also disjoint.

Since ϕ is continuous by the definition of a homomorphism, each of the sets $\phi^{-1}(C_0)(b)$ is compact, as the inverse image of a compact set under a continuous map from a compact space is compact. Hence there is a $\delta > 0$ such that the distance between any two different sets $\phi^{-1}(C_0(b))$ is $> \delta$. Choose n with $2^{-n} < \delta$. Let $x, x' \in X$. Then

$$[x_{-n}, x_n] = [x'_{-n}, x'_n] \Rightarrow \phi(x)_0 = \phi(x')_0$$

since then x and x' are at distance at most 2^{-n} and hence must lie in the same $\phi^{-1}(C_0(b))$. In other words, there is a map

$$\Phi : \mathcal{A}^{2n+1} \to \mathcal{B}$$

such that

$$\phi(x)_0 = \Phi([x_{-n}, x_n]).$$

But now the condition that $\sigma_Y \circ \phi = \phi \circ \sigma_X$ implies that

$$\phi(x)_1 = \Phi([x_{-n+1}, x_{n+1}])$$

and more generally that

$$\phi(x)_j = \Phi([x_{j-n}, x_{j+n}]). \tag{12.1}$$

Such a map is called a **sliding block code** of block size $2n+1$ (or "with memory n and anticipation n") for obvious reasons.

Conversely, suppose that ϕ is a sliding block code. It clearly commutes with the shifts. If x and x' agree on a central block of size $2N+1$, then $\phi(x)$ and $\phi(y)$ agree on a central block of size $2(N-n)+1$. This shows that ϕ is continuous. In short, we have proved

Proposition 12.1.1. *A map ϕ between two shift dynamical systems is a homomorphism if and only if it is a sliding block code.*

The advantage of this proposition is that it converts a topological property, continuity, into a finite type property - the sliding block code. Conversely, we can use some topology of compact sets to derive facts about sliding block codes.

For example, it is easy to check that a bijective continuous map $\phi : X \to Y$ between compact metric spaces is a homeomorphism, i.e. that ϕ^{-1} is continuous. Indeed, if not, we could find a sequence of points $y_i \in Y$ with $y_n \to y$ and $x_n = \phi^{-1}(y_k) \not\to x = \phi^{-1}(y)$. Since X is compact, we can find a subsequence of the x_n which converge to some point $x' \neq x$. Continuity demands that $\phi(x') = y = \phi(x)$ and this contradicts the bijectivity. From this we conclude that the inverse of a bijective sliding block code is continuous, hence itself a sliding block code - a fact that is not obvious from the definitions.

12.2 Shifts of finite type.

Let M be any positive integer and suppose that we map $X_{\mathcal{F}} \subset \mathcal{A}^{\mathbb{Z}}$ into $(\mathcal{A}^M)^{\mathbb{Z}}$ as follows: A "letter" in $\mathcal{A}^M$ is an M-tuplet of letters of $\mathcal{A}$. Define the map $\phi : X_{\mathcal{F}} \to (\mathcal{A}^M)^{\mathbb{Z}}$ by letting $\phi(x)_i = [x_i, x_{i+M}]$. For example, if $M = 5$ and we write the 5-tuplets as column vectors, the element x is mapped to

$$\ldots, \begin{pmatrix} x_{-1} \\ x_0 \\ x_1 \\ x_2 \\ x_3 \end{pmatrix}, \begin{pmatrix} x_0 \\ x_1 \\ x_2 \\ x_3 \\ x_4 \end{pmatrix}, \begin{pmatrix} x_1 \\ x_2 \\ x_3 \\ x_4 \\ x_5 \end{pmatrix}, \begin{pmatrix} x_2 \\ x_3 \\ x_4 \\ x_5 \\ x_6 \end{pmatrix}, \ldots.$$

This map is clearly a sliding block code, hence continuous, and commutes with shift hence is a homomorphism. On the other hand it is clearly bijective since we can recover x from its image by reading the top row. Hence it is a conjugacy of X onto its image. Call this image X^M.

We say that X is of **finite type** if we can choose a finite set $\mathcal{F}$ of forbidden words so that $X = X_{\mathcal{F}}$.

12.2.1 One step shifts.

If w is a forbidden word for X, then any word which contains w as a substring is also forbidden. If $M+1$ denotes the largest length of a word in $\mathcal{F}$, we may enlarge all the remaining words by adding all suffixes and prefixes to get words of length $M+1$. Hence, with no loss of generality, we may assume that all the words of $\mathcal{F}$ have length $M+1$. So $\mathcal{F} \subset \mathcal{A}^{M+1}$. Such a shift is called an M-step shift. But if we pass from X to X^{M+1}, the elements of $(\mathcal{A})^{M+1}$ are now the alphabet. So excluding the elements of $\mathcal{F}$ means that we have replaced the alphabet $\mathcal{A}^{M+1}$ by the smaller alphabet $\mathcal{E}$, the complement of $\mathcal{F}$ in $\mathcal{A}^{M+1}$. Thus $X^{M+1} \subset \mathcal{E}^{\mathbb{Z}}$. The condition that an element of $\mathcal{E}^{\mathbb{Z}}$ actually belong to X is easy to describe: An $(M+1)$-tuplet y_i can be followed by an $(M+1)$-tuplet y_{i+1} if and only if the last M entries in y_i coincide with the first M entries in y_{i+1}. All words $w = yy'$ which do not satisfy this condition are excluded. All these words have length two. We have proved that

the study of shifts of finite type is the same as the study of one step shifts.

12.3 Directed multigraphs.

We can rephrase the above argument in the language of graphs. For any shift and any positive integer K and a shift X of finite type we let $\mathcal{W}_K(X)$ denote the set of all admissible words of length K. Suppose that X is an M-step shift. Let us set

$$\mathcal{V} := \mathcal{W}_M(X),$$

and define

$$\mathcal{E} = \mathcal{W}_{M+1}(X)$$

as before.

Define maps

$$i : \mathcal{E} \to \mathcal{V}\,, \quad t : \mathcal{E} \to \mathcal{V}$$

to be

$$i(a_0a_1 \cdots a_M) = a_0a_1 \cdots a_{M-1} \quad t(a_0a_1 \cdots a_M) = a_1 \cdots a_M.$$

Then a sequence $u = \cdots u_1u_0u_1u_2 \cdots \in \mathcal{E}^{\mathbf{Z}}$, where $u_i \in \mathcal{E}$ lies in X^{M+1} if and only if

$$t(u_j) = i(u_{j+1}) \tag{12.2}$$

for all j.

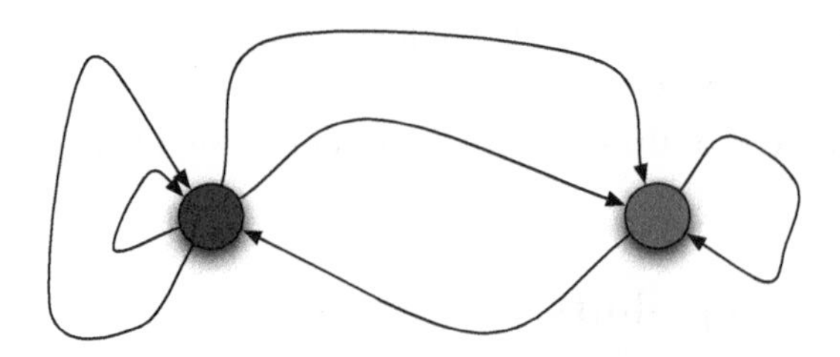

Figure 12.1: A directed multigraph with two vertices and six edges.

So let us define a **directed multigraph** (**DMG** for short) G to consist of a pair of sets $(\mathcal{V}, \mathcal{E})$ (called the set of vertices and the set of edges) together with a pair of maps

$$i : \mathcal{E} \to \mathcal{V}, \quad t : \mathcal{E} \to \mathcal{V}.$$

We may think the edges as joining one vertex to another, the edge e going from $i(e)$ (the initial vertex) to $t(e)$ the terminal vertex. The edges are "oriented" in the sense each has an initial and a terminal point. We use the phrase "multi"graph since nothing prevents several edges from joining the same pair of vertices. Also we allow for the possibility that $i(e) = t(e)$, i.e. for "loops". Starting from any **DMG** G, we define $Y_G \subset \mathcal{E}^{\mathbb{Z}}$ to consist of those sequences for which

$$t(u_j) = i(u_{j+1}) \qquad (12.2)$$

holds. This is clearly a step one shift.

We have proved that any shift of finite type is conjugate to Y_G for some **DMG** G.

12.3.1 The adjacency matrix of a directed multigraph.

Suppose we are given $\mathcal{V}$. Up to renaming the edges which merely changes the description of the alphabet, $\mathcal{E}$, we know G once we know how many edges go from i to j for every pair of elements $i, j \in \mathcal{V}$. This is a non-negative integer, and the matrix

$$A = A(G) = (a_{ij})$$

is called the **adjacency matrix** of G. The adjacency matrix f the graph in Figures 12.1 and 12.2 is

$$\begin{pmatrix} 2 & 2 \\ 1 & 1 \end{pmatrix}.$$

All possible information about G, and hence about Y_G is encoded in the matrix A. Our immediate job will be to extract some examples of very useful properties of Y_G from algebraic or analytic properties of A. In any event, we have reduced the study of finite shifts to the study of square matrices with non-negative integer entries.

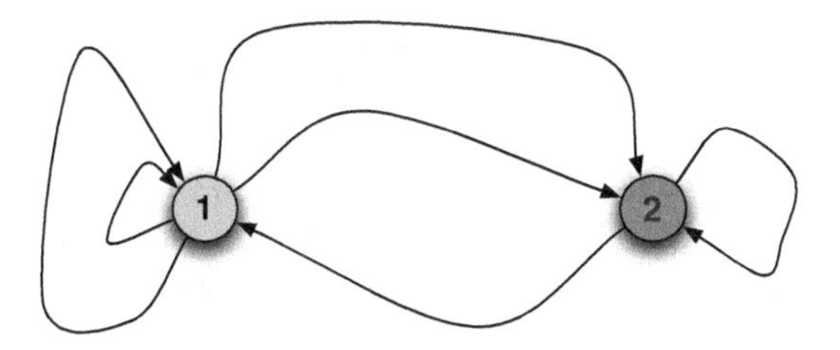

Figure 12.2: The preceding graph with the vertices labeled.

12.3.2 The number of fixed points.

For any dynamical system, (M, F) let $p_n(F)$ denote the number (possibly infinite) of fixed points of F^n. These are also called periodic points of period n. We shall show that if A is the adjacency matrix of the **DMG** G, and (Y_G, σ_Y) is the associated shift, then

$$p_n(\sigma_Y) = \operatorname{tr} A^n. \tag{12.3}$$

To see this, observe that for any vertices i and j, a_{ij} denotes the number of edges joining i to j. Squaring the matrix A, the ij component of A^2 is

$$\sum_k a_{ik}a_{kj}$$

which is precisely the number of words (or paths) of length two which start at i and end at j. By induction, the number of paths of length n which join i to j is the ij component of A^n. Hence the ii component of A^n is the number of paths of length n which start and end at i. Summing over all vertices, we see that $\operatorname{tr} A^n$ is the number of all cycles of length n. But if c is a cycle of length n, then the infinite sequence $y = \cdots ccccc \cdots$ is periodic with period n under the shift. Conversely, if y is periodic of period n, then $c = [y_0, y_{n-1}]$ is a cycle of length n with $y = \cdots ccccc \cdots$. Thus $p_n(\sigma_Y) =$ the number of cycles of length $n = \operatorname{tr} A^n$. $\square$

12.3.3 The zeta function.

Let (M, F) be a dynamical system for which $p_n(F) < \infty$ for all n. A convenient bookkeeping device for storing all the numbers $p_n(F)$ is the **zeta function**

$$\zeta_F(t) := \exp\left(\sum_n p_n(F)\frac{t^n}{n}\right).$$

This was introduced and studied by Artin and Mazur in 1965 -"On periodic points",*Annals of Mathematics. Second Series* **81** pp. 8299.

At the moment, ζ_F is to be regarded as a formal power series in t. We shall soon find an "Euler product" formula for ζ_F where the prime numbers are replaced by orbits:

Let x be a periodic point (of some period) and let $m = m(x)$ be the minimum period of x. Let $\gamma = \gamma(x) = \{x, Fx, \ldots, F^{m-1}x\}$ be the orbit of x under F and all its powers. So $m = m(\gamma) = m(x)$ is the number of elements of γ. The number of elements of period n which correspond to elements of γ is m if $m|n$ and zero otherwise. If we denote this number by $p_n(F,\gamma)$ then

$$\exp\left(\sum_n p_n(F,\gamma)\frac{t^n}{n}\right) = \exp\left(\sum_j m\frac{t^{mj}}{mj}\right) =$$

$$\exp\left(\sum_j \frac{t^{mj}}{j}\right) = \exp\left(-\log(1-t^m)\right) = \frac{1}{1-t^m}.$$

Now

$$p_n(F) = \sum_\gamma p_n(F,\gamma)$$

since a point of period n must belong to some periodic orbit. Since the exponential of a sum is the product of the exponentials we conclude that

$$\zeta_F(t) = \prod_\gamma \left(\frac{1}{1-t^{m(\gamma)}}\right).$$

This is the "Euler product" mentioned above.

The zeta function of a directed multipgraph.

Let us specialize to the case (Y_G, σ_Y) for some **DMG**, G. We claim that

$$\zeta_\sigma(t) = \frac{1}{\det(I-tA)}. \tag{12.4}$$

Indeed,

$$p_n(\sigma) = \operatorname{tr} A^n = \sum \lambda_i^n$$

where the sum is over all the eigenvalues (counted with multiplicity). Hence

$$\zeta_\sigma(t) = \prod \exp\sum \frac{(\lambda_i t)^n}{n} = \prod\left(\frac{1}{1-\lambda_i t}\right) = \frac{1}{\det(I-tA)}. \quad \square$$

12.4 Topological entropy.

Let X be a shift space, and let $\mathcal{W}_n(X)$ denote the set of words of length n which appear in X. Let $w_n = \#(\mathcal{W}_n(X)$ denote the number of words of length n. Clearly $w_n \geq 1$ (as we assume that X is not empty), and

$$w_{m+n} \leq w_m \cdot w_n$$

and hence

$$\log_2(w_{m+n}) \leq \log_2(w_m) + \log_2(w_n).$$

This implies that

$$\lim_{n\to\infty} \frac{1}{n} \log_2 w_n$$

exists on account of the following:

Lemma 12.4.1. *Let $a_1, a_2 \dots$ be a sequence of non-negative real numbers satisfying*

$$a_{m+n} \leq a_m + a_n.$$

Then $\lim_{n\to\infty} \frac{1}{n} a_n$ exists and in fact

$$\lim_{n\to\infty} \frac{1}{n} a_n = \lim_{n\to\infty} \inf_{k\geq n} \frac{1}{k} a_k.$$

Proof. Set $a := \lim_{n\to\infty} \inf_{k\geq n} \frac{1}{k} a_k$. Since $\frac{a_{rk}}{rk} \leq \frac{a_r}{r}$ by the hypothesis of the lemma, we see that the non-decreasing sequence defining a is bounded, so a is finite. For any $\epsilon > 0$ we must show that there exists an $N = N(\epsilon)$ such that

$$\frac{1}{n} a_n \leq a + \epsilon \quad \forall\, n \geq N(\epsilon).$$

Choose some integer r such that

$$a_r < a + \frac{1}{2}\epsilon.$$

Such an $r \geq 1$ exists by the definition of a. Using the inequality $a_{m+n} \leq a_m + a_n$, we get, if $0 \leq j < r$

$$\frac{a_{mr+j}}{mr+j} \leq \frac{a_{mr}}{mr+j} + \frac{a_j}{mr+j}.$$

Decreasing the denominator the right hand side is $\leq$

$$\frac{a_{mr}}{mr} + \frac{a_j}{mr}.$$

There are only finitely many a_j which occur in the second term, and hence by choosing m large we can arrange that the second term is always $< \frac{1}{2}\epsilon$. Repeated application of the inequality in the lemma gives

$$\frac{a_{mr}}{mr} \leq \frac{m a_r}{mr} = \frac{a_r}{r} < a + \frac{1}{2}\epsilon.$$

□

Thus we define

$$h(X) = \lim_{n\to\infty} \frac{1}{n} \log_2 \#(\mathcal{W}_n(X)), \tag{12.5}$$

and call $h(X)$ the **topological entropy** of X. (This is a standard but unfortunate terminology, as the topological entropy is only loosely related to the concept of entropy in thermodynamics, statistical mechanics or information theory as discussed in the preceding chpater). To show that it is an invariant of X we prove

Theorem 12.4.1. *Let $\phi : X \to Y$ be a factor (i.e. a surjective homomorphism). Then $h(Y) \leq h(X)$. In particular, if h is a conjugacy, then $h(X) = h(Y)$.*

Proof. We know that ϕ is given by a sliding block code, say of size $2m+1$. Then every block in Y of size n is the image of a block in X of size $n + 2m + 1$, i.e.

$$\frac{1}{n}\log_2 \#(\mathcal{W}_n(Y)) \leq \frac{1}{n}\log_2 \#(\mathcal{W}_{n+2m+1}(X)).$$

Hence

$$\frac{1}{n}\log_2 \#(\mathcal{W}_n(Y)) \leq \left(\frac{n+2m+1}{n}\right)\frac{1}{n+2m+1}\log_2 \#(\mathcal{W}_{n+2m+1}(X)).$$

The expression in parenthesis tends to 1 as $n \to \infty$ proving that $h(Y) \leq h(X)$. If ϕ is a conjugacy, the reverse inequality applies. □

The topological entropy and the adjacency matrix.

The adjacency matrix of a **DMG** has non-negative integer entries, in particular non-negative entries. If a row consisted entirely of zeros, then no edge would emanate from the corresponding vertex, so this vertex would make no contribution to the corresponding shift. Similarly if column consisted entirely of zeros. So without loss of generality, we may restrict ourselves to graphs whose adjacency matrix contains at least one positive entry in each row and in each column. This implies that if A^k has *all* its entries positive, then so does A^{k+1} and hence all higher powers. Recall that a matrix with non-negative entries which has this property is called **primitive**. Also recall that a matrix with non-negative entries is called **irreducible** if for any ij there is some power n (depending on i and j) such that $(A^n)_{ij} \neq 0$. In terms of the graph G, the condition of being primitive means that for all sufficiently large n any vertices i and j can be joined by a path of length n. The slightly weaker condition of irreducibility asserts that for any i and j there exist $n = n(i,j))$ and a path of length n joining i and j.

Finally, recall the **Perron-Frobenius Theorem** which asserts every non-negative irreducible matrix A has a positive eigenvalue λ_A such that $\lambda_A \geq |\mu|$ for any other eigenvalue μ and also that $Av = \lambda_A v$ for some vector v all of whose entries are positive, and that no other eigenvalue has an eigenvector with all positive entries. We will use this theorem to prove:

Theorem 12.4.2. *Let G be a* **DMG** *whose adjacency matrix $A(G)$is irreducible. Let Y_G be the corresponding shift space. then*

$$h(Y_G) = \lambda_{A(G)}. \tag{12.6}$$

Proof. The number of words of length n which join the vertex i to the vertex j is the ij entry of A^n where $A = A(G)$. Hence

$$\#(\mathcal{W}_n(Y_G)) = \sum_{ij}(A^n)_{ij}.$$

Let v be an eigenvector of A with all positive entries, and let $m > 0$ be the minimum of these entries and M the maximum. Also let us write λ for λ_A. We have $A^n v = \lambda^n v$, or written out

$$\sum_j (A^n)_{ij} v_j = \lambda^n v_i.$$

Hence

$$m \sum_j (A^n)_{ij} \leq \lambda^n M.$$

Summing over i gives $m\#(\mathcal{W}_n(Y_G)) \leq rM\lambda^n$ where r is the size of the matrix A. Hence

$$\log_2 m + \log_2 \#(\mathcal{W}_n(Y_G)) \leq \log_2(Mr) + n\log_2 \lambda.$$

Dividing by n and passing to the limit shows that

$$h(Y_G) \leq \lambda_A.$$

On the other hand, for any i we have

$$m\lambda^n \leq \lambda^n v_i \leq \sum_j (A^n)_{ij} v_j \leq M \sum_j (A^n)_{ij}.$$

Summing over i gives

$$rm\lambda^n \leq M\#(\mathcal{W}_n(Y_G)).$$

Again, taking logarithms and dividing by n proves the reverse inequality

$$h(Y_G) \geq \lambda_A.$$

□

The Fibonacci example.

For example, if

$$A = \begin{pmatrix} 1 & 1 \\ 1 & 0 \end{pmatrix}$$

then

$$A^2 = \begin{pmatrix} 2 & 1 \\ 1 & 1 \end{pmatrix}$$

so A is primitive. Its eigenvalues are

$$\frac{1 \pm \sqrt{5}}{2}$$

so that

$$h(Y_G) = \frac{1+\sqrt{5}}{2}.$$

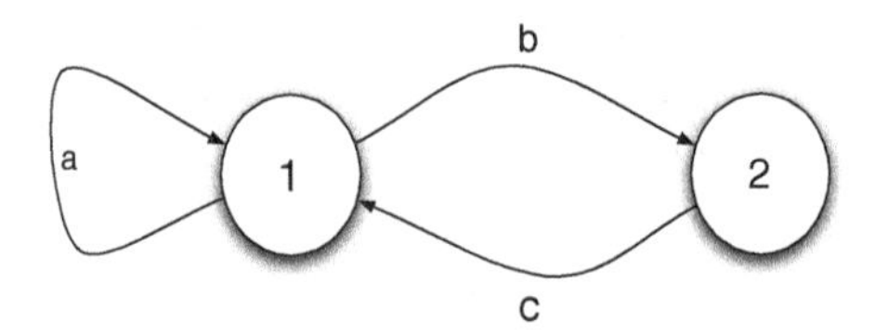

Figure 12.3: The graph of the Fibonacci shift.

12.5 Factors of finite shifts.

Suppose that X is a shift of finite type and $\phi : X \to Z$ is a surjective homomorphism, i.e. a factor. Then Z need not be of finite type. Here is an illustrative example. Let $\mathcal{A} = \{0, 1\}$ and let $Z \subset \mathcal{A}^{\mathbb{Z}}$ consist of all infinite sequences such that there are always an even number of zeros between any two ones. So the excluded words are

$$101,\ 10001,\ 1000001,\ 100000001, \ldots$$

(and all words containing them as substrings). It is clear that this can not be replaced by any finite list, since none of the above words is a substring of any other.

On the other hand, let G be the **DMG** associated with the matrix

$$A = \begin{pmatrix} 1 & 1 \\ 1 & 0 \end{pmatrix},$$

and let Y_G be the corresponding shift. We claim that there is a surjective homomorphism $\phi : Y_G \to Z$. To see this, assume that we have labelled the vertices of G as $1, 2$, that we let a denote the edge joining 1 to itself, b the edge joining 1 to 2, and c the edge joining 2 to 1. So the alphabet of the graph Y_G is $\{a, b, c\}$ and the excluded words are

$$ac\ bb, ba, cc$$

and all words which contain these as substrings. So if the word ab occurs in an element of Y_G it must be followed by a c and then by a sequence of bc's until the next a. Now consider the sliding block code of size 1 given by

$$\Phi :\ a \mapsto 1,\ b \mapsto 0,\ c \mapsto 0.$$

From the above description it is clear that the corresponding homomorphism is surjective.

We can describe the above procedure as assigning "labels" to each of the edges of the graph G; we assign the label 1 to the edge a and the label 0 to the edges b and c.

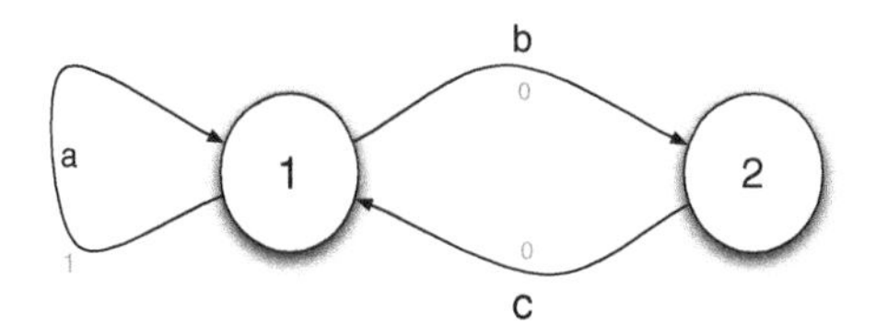

Figure 12.4: The graph of the Fibonacci shift, labeled.

It is clear that this procedure is pretty general: a **labeling** of a directed multigraph is a map:Φ: $\mathcal{E} \rightarrow \mathcal{A}$ from the set of edges of G into an alphabet $\mathcal{A}$. It is clear that Φ induces a homomorphism ϕ of Y_G onto some subshift of $Z \subset \mathcal{A}^{\mathbb{Z}}$ which is then, by construction a factor of a shift of finite type.

Conversely, suppose X is a shift of finite type and $\phi : X \rightarrow Z$ is a surjective homomorphism. Then ϕ comes from some sliding block code. Replacing X by X^N where N is sufficiently large we may assume that X^N is one step and that the block size of Φ is one. Hence we may assume that $X = Y_G$ for some G and that Φ corresponds to a labeling of the edges of G. We will use the symbol (G, L) to denote a **DMG** together with a labeling of its edges. We shall denote the associated shift space by $Y_{(G,L)}$.

Unfortunately, the term **sofic** is used to describe a shift arising in this way,i.e. a factor of a shift of finite type. (The term is a melange of the modern Hebrew mathematical term *sofi* meaning finite with an English sounding suffix.)

12.6 The Henon map and symbolic dynamics.

The Henon map on the plane (depending on two parameters b and c) is defined by

$$H = H_{b,c} : \ (x, y) \mapsto (x^2 - c - by, x).$$

The goal of this section is to show that for a suitable range of the parameters b and c, there is a subset $K = K(b, c) \subset \mathbb{R}^2$ which is invariant under the Henon map, and such that the restriction of H to K is conjugate to the full shift map on two letters.

This is a theorem of Devaney and Nitecki "Shift automorphisms in the Hénon mapping" *Communications in Mathematical Physics***67** (1979) pp. 137 - 146. It is a special case of a theorem of Knill "Topological entropy of standard type monotone twist maps" *Transactions of the American Mathematical Society* **348**, (1996) pp. 2999 - 3013.

We follow the treatment in [Knill].

A change of variables.

$$H = H_{b,c} : \ (x, y) \mapsto (x^2 - c - by, x).$$

Change variables

$$(q,p) := a(x,y) \quad a := \frac{1}{\sqrt{c}}.$$

In these variables the map is

$$T = T_{a,b}: \quad (q,p) \mapsto \left(\frac{q^2-1}{a} - bp, q\right).$$

If we write $\begin{pmatrix} q_n \\ q_{n-1} \end{pmatrix}$ for the n-th iterate then

$$\begin{pmatrix} q_{n+1} \\ q_n \end{pmatrix} = \begin{pmatrix} \frac{q_n^2-1}{a} - bq_{n-1} \\ q_n \end{pmatrix}$$

yielding the recurrence relation

$$aq_{n+1} + abq_{n-1} = q_n^2 - 1.$$

Let

$$X = \{-1,1\}^{\mathbb{Z}}$$

and let S be the shift map on X.

We look for a continuous map $q : X \to \mathbb{R}$ such that $q_n = q(S^n x),\ x \in X$ is a solution of the above recurrence relation.

Our problem as an implicit function problem.

Let $C(X)$ denote the space of continuous real valued functions on X, and let $F : \mathbb{R} \times C(X) \to C(X)$ be defined by

$$F(a,q)(x) : a \cdot q((Sx)) + ab \cdot q(S^{-1}x) - (q(x)^2 - 1).$$

We would like to find q such that

$$F(a,q) = 0.$$

For $a = 0$ a function q such that $|q| \equiv 1$ is a solution.

The partial derivative of F in the $C(X)$ direction is

$$F_q(a,q)(u) = a \cdot u \circ S + ab \cdot u \circ S^{-1} - 2q \cdot u.$$

At $a = 0$ we have

$$F_q(0,q)u = -2q \cdot u$$

which is an invertible map. Indeed since $|q| \equiv 1$, dividing by $-2q$ gives the inverse, and this inverse is a bounded linear map. We shall fix a solution by choosing $q(x) \equiv x_0$. The implicit function theorem says that for sufficiently small a we can find a $q = q_a$ depending continuously on a, such that

$$F(a, q_a) \equiv 0$$

and $q_0(x) \equiv x_0$.

We now use this solution q_a to construct a conjugacy between S acting on X and T acting on an invariant closed subset of $\mathbb{R}^2$. In what follows we will write q instead of q_a so as not to clutter up the formulas.

The conjugacy.

Define $\phi = \phi_a : X \to \mathbb{R}^2$ by

$$\phi(x) = \begin{pmatrix} q(x) \\ q(S^{-1}x) \end{pmatrix}.$$

ϕ is continuous, because q and S^{-1} are continuous. Using $F(a, q) = 0$, we check that

$$(\phi \circ S)(x) = \begin{pmatrix} q(Sx) \\ q(x) \end{pmatrix} = \begin{pmatrix} \frac{1}{a}(q(x)^2 - 1) - bq(S^{-1}x) \\ q(x) \end{pmatrix}$$

$$= T \begin{pmatrix} q(x) \\ q(S^{-1}x) \end{pmatrix} = (T \circ \phi)(x).$$

The map ϕ is injective because if two points x, y are mapped into the same point in $\mathbb{R}^2$ then the fact that $q_a(x)$ is close to x_0 and $q_a(y)$ is close to y_0 implies that $x_0 = y_0$. Then the conjugacy $(\phi \circ S^n)(x) = T^n \circ \phi(x)$ and the fact that T is a homeomorphism implies that $x_n = y_n$ for all n, i.e. $x = y$.

ϕ has a continuous inverse because every continuous bijective map from a compact space to a compact space has a continuous inverse. So the map ϕ is indeed a homeomorphism from X to a closed subset $K = \phi(X) \subset \mathbb{R}^2$ such that K is invariant under T and

$$\phi \circ S = T \circ \phi.$$

Bibliography

[Caswell] Hal Caswell. *Matrix Poulation Models, Construction, Analysis and Interpretation.* 2nd Edition, Sinauer Assoc., Inc., Publishers (2001)

[Devaney(1989)] Robert L. Devaney. *An introduction to chaotic dynamical systems.* Addison Wesley. (1989)

[Devaney(1992)] Robert L. Devaney. *A first course in dynamical systems.* Addison Wesley. (1992)

[Edgar] Gerald A. Edgar. *Measure, Topology, and Fractal Geometry.* Springer (1990)

[Ellner and Guckenheimer] Stephen P. Ellner and John Guckenheimer. *Dynamic Models in Biology.* Princeton University Press. (2002)

[Feller] W. Feller. *An Introduction to Probability Theory and its Applications, I.* Wiley (1950)

[Gulick] Denny Gulick. *Encounters with chaos.* McGraw Hill. (1992)

[Hofbauer and Sigmund] Josef Hofbauer and Karl Sigmund. *Evolutionary Games and Population Dynamics.* Cambridge University Press. (1998)

[Jaynes] E. T. Jaynes Prior probabilities. IEEE Transactions Systems Science and Cybernetics. , **SSC-4** 227-241 (1968)

[Knill] Oliver Knill *Topics in dynamical systems.* Course notes (1994) available on the web.

[Kuznetsov] Yu. A. Kuznetsov. *Elements of applied Bifurcation Theory.* Applied Mathematical Sciences **112** Springer. (1995)

[Langville and Meyer] Amy N. Langville and Carl D. Meyer. *Google's PageRank and Beyond.* Princeton University Press (2006)

[Lind and Marcus] Douglas Lind and Brian Marcus. *Symbolic Dynamics and Coding.* Cambridge University Press. (1995)

[Lynch] Stephen Lynch. *Dynamical Systems with Applications using MATLAB.* Birckhäuser. (2004)

[McMullen] Curtis T. McMullen. *Complex Dynamics and Renormalization.* Annals of Mathematics Studies **135** Princeton University Press (1994)

[Nelson] Edward Nelson. *Topics in Dynamics I: Flows.* Princeton University Press and University of Tokyo Press (1970). Available on the web.

[Robinson] Clark Robinson. *Dynamical Systems.* CRC Press (1995)

[Schroeder] Manfred Schroeder. *Fractals, Chaos, and Power Laws.* Freeman. (1990)

[Shub] Michael Shub. *Global Stability of Dynamical Systems.* Springer-Verlag (1987)

[Strogatz] Steven H. Strogatz. *Nonlinear Dynamics and Chaos.* Perseus Books (1994)

[Welsh] Dominic Welsh *Codes and Cryptography.* Oxford University Press (1987)

Index